"The authors identify several maj[...] [...]mic, including digitalization, passive [...] ts in consumption patterns. This thoughtful discu[...] [...] is for any long-term academic and policy discourse regarding the qu[...]ion on where we should go once the pandemic subsides."

—**Eran Feitelson,** *Leon J. and Alyce K. Ell*
Chair in Environmental Studies, The Hebrew University of
Jerusalem, Israel

"The world is rapidly changing, calling for a change in our assumptions, lifestyles, and the way we consume. This book brings together the knowledge of well-renowned scholars and practitioners from different parts of the world, and presents a global overview of our current situation. It offers tentative solutions and identifies the importance of personal engagement for addressing the key causes in favor of both the human race and the planet. It is a must-read for those who want to be part of this transformation"

—**Marcus Nakagawa,** *São Paulo, Brazil; Founder of ABRAPS*
(Brazilian Association of Sustainability Consultants and Professionals),
TEDx Talk speaker and marketing sciences professor

"This is a timely and fascinating book written during the pandemic. Many of the impacts of Covid-19 upon human societies have yet to be discerned, but some have become clear. The authors of this book have clearly demonstrated the accelerated processes caused by the pandemic, i.e., digitalization, localization and community connectivity, mindful consumption, and sharing and counteracting waste. The book is thoughtful and witty, full of insights and inspirations. It is one of the first remarkable academic responses to the effects brought about by Covid-19, which will certainly contribute to deepening our understanding of the pandemic's social and cultural impacts."

—**Ning Wang,** *Professor of Department of Sociology and Social Work,*
School of Sociology and Anthropology, Sun Yat-sen University,
Guangzhou, China

"The impact of Covid-19 is a central concern to futures-oriented thinkers and actors. This book, written by a diverse community of scholars across the world, offers a very early, but comprehensive and thoughtful, glimpse of what these might be, organized to be accessible and useful to policy makers and academics."

—**John R. Ehrenfeld,** *Author: The Right Way to Flourish:*
Reconnecting with the Real World (MIT, retired)

"This book, written in the midst of the pandemic, provides interesting analyses on trends and scenarios, and lays the ground for more in-depth research on how to ensure that some of the changes will last, how to turn our leaders into responsible managers (there is none so far), how to lay long-lasting foundations for a transition to responsible consumption and production, and sustainable lifestyles, wisely using public funds for long-term objectives such as climate change and sustainability, and not just in emergencies of wars and pandemics. Beyond stimulating dialogues, it is hoped that this book will stimulate joint action for a global response."

—**Arab Hoballah,** *Team Leader, Switch-Asia SCP Facility,*
Ex-Chief Sustainable Consumption and Production,
UN Environmental Program

"This book is a compelling dive into the implications of the unfolding coronavirus pandemic and an exploration of its long-term effects on society and our lifestyles. In tracing these patterns across several countries, the authors note the striking similarities in the impacts on our daily lives including from an increasingly virtual world. They reveal conflicting evidence that these trends are leading to more sustainability; however, the authors conclude that there is a key role for policymakers in enabling a transition that supports lifestyles grounded in health, justice, and planetary and community wellbeing."

—**Vanessa Timmer,** *Executive Director, OneEarth, Vancouver,*
Canada; Senior Research Fellow, Utrecht University, The Netherlands

Sustainable Lifestyles after Covid-19

This book takes an in-depth look at Covid-19-generated societal trends and develops scenarios for possible future directions of urban lifestyles.

Drawing on examples from Brazil, China, and Israel, and with a particular focus on cities, this book explores the short- and long-term changes in individual consumers and citizen behavior as a result of the Covid-19 pandemic. On the basis of extensive market and opinion research data, aggregate data, observational evidence, and news reports, the authors provide a detailed account of the transformations that have occurred as a result of a triple shock of public health emergency, economic shutdown, and social isolation. They also examine which of these behavioral changes are likely to become permanent and consider whether this may ultimately promote or restrain sustainable lifestyle choices.

Innovative and timely, this book will be of great interest to students, scholars, and professionals researching and working in the areas of sustainable consumption, urban and land use planning, and public health.

Fabián Echegaray is the Founder and Managing Director of Market Analysis, Florianópolis, SC, Brazil.

Valerie Brachya is Senior Associate Researcher at The Jerusalem Institute for Policy Research and Lecturer in Sustainability Policy at the Porter School of the Environment and Earth Sciences, Tel Aviv University, Israel.

Philip J. Vergragt is Emeritus Professor at TU Delft, The Netherlands; Research Professor at Clark University, USA; and a co-founder of SCORAI.

Lei Zhang is Associate Professor at the School of Environment and Natural Resources at Renmin University of China, Beijing, China.

Routledge – SCORAI Studies in Sustainable Consumption
Series Editors:
Halina Szejnwald Brown, *Professor Emerita at Clark University, USA.*
Philip J. Vergragt, *Emeritus Professor at TU Delft, The Netherlands; Research Professor at Clark University, USA.*
Lucie Middlemiss, *Associate Professor and Co-Director of the Sustainability Research Institute, Leeds University, UK.*
Daniel Fischer, *Associate Professor for Consumer Communication and Sustainability, Wageningen Research and University, The Netherlands.*

This series aims to advance conceptual and empirical contributions to this new and important field of study. For more information about The Sustainable Consumption Research and Action Initiative (SCORAI) and its activities please visit www.scorai.org.

Social Innocation and Sustainable Consumption
Research and Action for Societal Transformation
Edited by Julia Backhaus, Audley Genus, Sylvia Lorek, Edina Vadovics and Julia M. Wittmayer

Power and Politics in Sustainable Consumption Research and Practice
Edited by Cindy Isenhour, Mari Martiskainen and Lucie Middlemiss

Local Consumption and Global Environmental Impacts
Accounting, Trade-offs and Sustainability
Kuishuang Feng, Klaus Hubacek and Yang Yu

Subsistence Agriculture in the US
Reconnecting to Work, Nature and Community
Ashley Colby

Sustainable Lifestyles after Covid-19
Fabián Echegaray, Valerie Brachya, Philip J. Vergragt, and Lei Zhang

For more information about this series, please visit: www.routledge.com/ Routledge-SCORAI-Studies-in-Sustainable-Consumption/book-series/ RSSC

Sustainable Lifestyles after Covid-19

Fabián Echegaray, Valerie Brachya,
Philip J. Vergragt, and Lei Zhang

Routledge
Taylor & Francis Group

LONDON AND NEW YORK

First published 2021
by Routledge
2 Park Square, Milton Park, Abingdon, Oxon OX14 4RN

and by Routledge
605 Third Avenue, New York, NY 10158

Routledge is an imprint of the Taylor & Francis Group, an informa business

British Library Cataloguing-in-Publication Data
A catalogue record for this book is available from the British Library

Library of Congress Cataloging-in-Publication Data
A catalog record for this book has been requested

ISBN: 978-0-367-75409-9 (hbk)
ISBN: 978-0-367-75411-2 (pbk)
ISBN: 978-1-003-16239-1 (ebk)

Typeset in Times New Roman
by Apex CoVantage, LLC

Contents

Preface

The idea of writing a book emerged within the context of the SCORAI (Sustainable Consumption Research and Action Initiative; www.scorai.net) network, which brings together researchers and practitioners from a wide range of disciplines to exchange knowledge and views concerning sustainable consumption and to influence activists and policy makers.

The network is largely dominated by participants from North America and Europe, but has attracted over the years small groups from Brazil, Israel, and China as well as individuals from other countries and continents. Philip Vergragt of Boston, one of the founders of SCORAI, invited leading representatives of the international groups to meet online and exchange ideas. Little did he imagine that this group would not only present papers at the fourth SCORAI bi-annual conference in June 2020 but also continue together through the summer and autumn months of Covid-19 to write a book.

Fabian Echegaray from Brazil initiated writing a book together, and it turned into a fascinating and intellectually stimulating experience. As we talked together on Zoom, we realized that despite the huge differences in political regimes, economic structures, and social organization, we actually found very similar behavioral responses. Lei Zhang from China was the first to present what appeared to be long-term societal transformations when she remarked "everything has gone online". Valerie Brachya of Israel realized soon afterwards that that was exactly what was happening around her.

The authors put their experiences together in writing this book but each chapter was the responsibility of the lead author. The chapters are integrated into a common coherent framework, but each is written from the lead author's viewpoint and presents some details of their own country's experience.

The book was written during the summer and fall of 2020 when the pandemic was by no means over and actually resurging. The worldwide hope was that by mid-2021 a vaccine would enable the world to overcome the worst health effects, but it was clear that the economic and cultural effects would be much more long-lasting. The authors based their chapters on the limited availability of data at that time and therefore their assertions are made with a high level of uncertainty. No doubt with multiple research and hard evidence it will be possible to provide a firm basis for such or other assertions in the future – but that will be too late for policy makers. The authors therefore conclude by presenting diverse directions of strategy and policy that they hope will stimulate an ongoing dialogue.

The authors wish to thank Halina Brown for reading the entire earlier version of this manuscript and for proposing valuable suggestions and data that greatly improved the final version of this book.

1 Introduction

How Covid-19 reshaped lifestyles across societies

Pandemics are life-changing events. They revamp how society functions, the choices and practices of individuals, the role of authorities and institutions, and the values and priorities of a polity. Plagues in the 14th century, the Spanish flu in the early 20th century, and the Covid-19 outbreak in 2020 share common traits, similar to recent occurrences of more local epidemics. Present-day observers of major virus crises are inclined to over-emphasize short-term repercussions and overlook long-term changes. Sometimes it is difficult to separate the implications of concurrent events, such as the major long-term effects of the Spanish flu together with the aftermath of World War I. While we recognize the immediate devastating health and economic effects of the Covid-19 pandemic, this book focuses mainly on its potential long-term societal implications.

Covid-19 disrupted social and economic order across the world, largely as the result of regulatory or voluntary containment measures. Governments and communities restricted movement to discourage gatherings and to enable public health services to cope with needs within the capacity of their medical systems. Policies across the globe, following recommendations by the WHO, aimed to interrupt the transmission of the Covid-19 virus and focused on three approaches affecting core dimensions of life:

- Declaring a public health emergency.
- Imposing a shutdown on economic activities.
- Defining social isolation and physical distancing as the ultimate prophylaxis for individuals.

They were accompanied by other measures that intervened in the market economy, for example, to secure basic food supplies and to provide stimulus funds and a safety net to offset negative social and economic consequences of the virus.

Governmental interventions affected some three-quarters of world population almost simultaneously, cascading and magnifying their effects on a totally unanticipated scale and direction, such as the intensification of social inequality (The Economist, 2020).

Three dimensions: health, economy, social relations

The public health emergency measures required the improvement of clean water supply and sanitation, but they also generated a broader cultural transformation relating to social beliefs and norms around hygiene, wellbeing, and safety. The traditional notions of cleanliness were expanded from their customary meanings of aesthetic quality, freshness, or moral purity to additional meanings, such as safety, comfort, and wellbeing (Neves, 2004). Wellbeing became detached from affluence or subjective plenitude and was reframed as disinfection and defensive, health-centric self-protective routines. Studies connecting handwashing rituals with the speed of transmission of the virus lent further legitimacy to this cultural intervention (Progrebna and Kharlamov, 2020).

The economic dimension of counter-pandemic actions closed down activities that involved face-to-face contact; those that could transferred their mode of operation to virtual environments, such as remote working and e-commerce, did so whenever possible. In practice, the sudden economic shutdown imposed by governments disrupted commercial activities, brought the consumer economy to a sudden halt, and resulted in widespread bankruptcy and financial hardship. Consumption in terms of volumes, patterns, and procedures was redefined. Layoffs, furloughs, and reduced paychecks caused severe loss of income, while expenses and indebtedness increased substantially. This picture repeated itself across the globe, in both developed and developing countries, particularly where states constrained by fiscal shortages could not offer emergency support and bailout packages. Inequality increased between and within countries, depending on who lost income and on whether some form of social security was available.

In the social realm, patterns of socializing and interpersonal connectivity were redrawn. Social isolation and physical distancing frequently confined people to their homes, reduced face-to-face interpersonal interaction to a minimum, intensified interaction within the household, and propelled social connections to the internet-mediated online sphere.

Despite wide diversity in political, economic, and social structures, most countries around the world experienced surprisingly similar outcomes. Nearly all domains of life were affected, from working to studying, from entertainment to shopping, from socializing to family life and love relationships, from the meaning of household and living habitats to the perceptions and expectations of citizenship.

Implications for sustainability

The links between Covid-19-related trends and sustainability are indirect. The impacts relevant to sustainability are the result of changes in societal structures and individual lifestyles that are likely to continue after solutions are found to health and safety risks.

Sustainability is often conceptualized as comprising three pillars: environmental, social, and economic. The UN Sustainable Development Goals have generally been accepted as the world's agreement on what constitutes a holistic view of sustainability. All 17 UN Sustainable Development Goals were affected by the pandemic, some more than others. The sudden halt of economic activity and travel temporarily reduced greenhouse gas (GHG) emissions and thus could be considered as demonstrating how the world could move toward goal 13: mitigating climate change (Nakada and Urban, 2020; Markard and Rosenbloom, 2020). However, Covid-19 hit vulnerable groups, causing a reversal of the progress that had been made in reducing local, regional, and global poverty and increasing inequality, including gender inequality. The economy suffered a major loss across the globe with an average contraction of −4.4% (IMF, 2020). Even if economic outlooks point to a steeper decline in advanced economies, emerging markets of Latin America were among the worst hit by recession (IMF, 2020).

The focus on climate change led at first to an over-optimistic assessment of the impacts of Covid-19 on emissions of air pollutants, celebrating cleaner air and lower GHG emissions (LeQuéré et al., 2020), without due consideration to its social repercussions (Cohen, 2020). Clear skies above Tiananmen Square, the sights of Himalayas in Northern India once smog cleared, and incursion of wild animals into cities epitomized an environmental dividend brought by lockdown. However positive the ecological effect of Covid-19-driven economic shutdown and social containment, it cannot be disconnected from drastic mental, physical, and dietary consequences. There is nothing sustainable in greening consumption or reducing emissions if these are accompanied by skyrocketing rates of anxiety, alcoholism, highly processed food consumption that boosts overweight or obesity, sleep disorders, or internet-dependent compulsive behaviors (Banerjee and Rai, 2020; Brooks et al., 2020; Huang and Zhao, 2020; Torales et al., 2020). There is nothing sustainable when cleaner skies or habits of frugality due to lockdown come at the cost of permanent unemployment and higher financial pressures on families and households sliding into poverty or misery, or food and home scarcity. There is little to celebrate in environmental preservation outcomes from a quarantine that also increases female overburden with home chores, greater inequality against women as they bear the costs of furloughs and layoffs more than men (Power, 2020), a rise in domestic violence against women and children, and increasing social unrest and crime rates (Bradbury-Jones and Isham, 2020).

Many scholars proposed that social norms and practices will return to a "back-to-normal" scenario as soon as restrictions are lifted (Boons et al., 2020). That proposal was based on the assumption that the new conditions leading to altered practices during the pandemic did not become new societal norms. Indeed, loss of income may delay, but not eliminate, purchasing and it may not itself drive a long-term force for changing purchasing behavior.

The question is whether current changes, generated by the need to do things differently during the periods of lockdown, will persist beyond the pandemic. Previous crises affecting the state of the economy, such as the Katrina hurricane in the US, offer evidence of lasting lifestyle changes to cope with a downward mobility situation and to subjectively integrate simplified ways of living as a new normal (Kennett-Hensel et al., 2012). The more direct, radical, and substantively traumatic the experience, the longer lasting effects of priorities and lifestyle changes.

This book proposes that some of the changes induced by the pandemic are likely to continue, at least in part, when the pandemic is over and restrictions are removed. This is because the Covid-19-driven transformations have altered the conditions that influence lifestyles and shape attitudes and behaviors. The sudden acceleration of digital systems, the focus on essential rather than superfluous goods, the enjoyment of non-commercialized forms of leisure in outdoor parks, and different forms of status signaling may be here to stay.

This book proposes that the way by which implications of the pandemic on sustainability should be considered is whether lifestyles have become more sustainable. It does not start from the sustainability discourse, such as how changing consumption behavior could promote reductions in GHG emissions. Rather, it presents the trends as they seem to be evolving through the Covid-19 pandemic, considers how they are likely to affect the way people live their daily lives and the context in which they live, and then poses the question of whether that is likely to lead to more or less sustainable lifestyles.

Studying lifestyles goes beyond consumption practices. Lifestyles constitute "a more or less integrated set of practices which an individual embraces, not only because such practices fulfil utilitarian needs, but because they give material form to a particular narrative of self-identity" (Giddens, 1991: 80). We concur that lifestyle choices are an expression of "decisions not only about how to act, but who to be" (Giddens, 1991: 81).

Changes in the perception of needs and social identity, and a re-evaluation of ideals, expectations, and what is essential in life, may break previously habitual routines and patterns of behavior. The authors therefore propose that this may be "a moment of change" when some of the barriers that condition the context where unsustainable behaviors are locked in may have been removed or weakened and some opportunities for lifestyle changes may have opened up. Some of the changes precipitated by the pandemic may lead to more sustainable lifestyles but others may halt or decelerate the pre-pandemic trends that were in a positive direction, and may now be lost (as will be seen in Chapter 5 on Cities). At the same time, numerous surveys indicate a strong wish among individuals from different societies to change their personal lives in a more sustainable and equitable direction, particularly in Global South countries (WEF (World Economic Forum)-Ipsos survey, 2020; WWF (World Wildlife Fund)-GlobeScan survey, 2020). There is therefore a social momentum and an opportunity for change, termed by WEF as "The Great Reset".

This book follows an optimistic view and proposes that many changes in the right direction generated by Covid-19 could become irreversible, but some will need deliberate encouragement by governments and policy makers.

A wider geographical overview

Most research published in English on the effects of Covid-19 is based on data from Europe and North America (Boons et al., 2020; SSPP, 2020), countries where many characteristics of social composition, economic structure, and living conditions (including access to internet) are similar. This book adds data, viewpoints, and experiences from very different societies, in the Global South like Brazil, the Global East (China), and the Middle East (Israel).

These countries vary widely in their political regimes, from a one-party regime (China) to multiparty democratic regimes (Brazil and Israel). The degree of trust in government during the lockdown also diverges, from full compliance with enforcement (China), a low trust, decentralized management of restrictions on movement (Brazil) to a mixed central/local management with loss of trust (Israel). They vary widely in the structure of their populations and levels of urbanization. Israel has a dominantly urban population (nearly 93%); almost all the population has access to clean water and sewage collection and treatment. Almost half the working population could work from home during the lockdown, with easy access to internet. This is in contrast to countries of the Global South, such as Brazil, where at least 15% of households (affecting 31 million persons) lack a domestic sewage system and drinking water (IBGE, 2019), which was a major barrier to implementing increased hygiene and sanitation. A total of 11.5 million

Brazilians live in overcrowded dwellings in shantytowns and poor tenements, which could not implement social isolation (IBGE, 2019). Remote working was a very limited option when 40% of jobs are in the "informal" economy and 20% of households lack access to internet (IBGE, 2019).

China has been experiencing rapid urbanization and industrialization over the past four decades. By the end of 2019, nearly 61% of the population was urban, domestic consumption contributed 58% to GDP growth, and internet penetration reached 65% (CNNIC, 2020). China is the world's largest online retail market and the world's largest mobile payment market. Its digital economy accounts for 34.8% of GDP.

Chapters of the book

The book is organized into nine chapters. Following the Introduction, Chapters 2 and 3 present evolving trends that are affecting lifestyles during the pandemic and consider where they may take us in the future. The next three chapters consider the contexts in which lifestyles manifest themselves: the home, the community, and the city (Chapters 4, 5, and 6). These are followed by an analysis of whether the reconfigured lifestyles may lead to more sustainable consumption behaviors (Chapter 7) and how policies and institutions can influence whether they will be more sustainable (Chapter 8). The final chapter presents insights that derive from the study of lifestyles during the pandemic and presents questions for future dialogue (Chapter 9). Here follows a more detailed outline of the book.

> Chapter 2 presents how Covid-19 affected societal trends in relation to ten domains of practice, and their implications for promoting or disrupting sustainable lifestyles. Three situations are identified. Evolving trends that were *accelerated* by the pandemic, affecting more sectors of the population in deeper ways, for example, digitalization and mindful consumerism. The virus outbreak halted or reversed ongoing trends, which *decelerated*, such as the sharing economy. Lastly, public health emergency measures generated some *unanticipated developments*, which are likely to become permanent. Unanticipated changes included the role of the home, the concept of wellbeing, and forms of citizen engagement. The chapter presents a perspective as seen mainly from Brazil.
>
> Chapter 3 looks at scenarios of future lifestyles and what the next normal could look like at the individual level once restrictions are lifted. On the basis of scarcity theory, which allocates higher priority to resources and experiences that are in relatively short supply, this chapter proposes an analytical framework that focuses on two

domains highly affected by Covid-19: consumption behavior and social connectivity. Four scenarios are presented with orientations and behaviors across major domains of practice. It presents a perspective as seen from Brazil.

Chapter 4 focuses on the changing role of the home. Since the home stopped being merely the space where sleeping, washing, relaxing, and dressing takes place, it became the center stage for all other daily life activities, including working, purchasing, and exercising. Moreover, increased centering of life around the home affected the role of shopping as conveying social identity and status when lockdown conditions defeated the purpose of conspicuous consumption. Both time and space were compressed and family relationships were intensified, for better or for worse. The chapter presents a historical perspective of how attitudes to the role of the home have changed over time, both for utilitarian purposes and as a social context. It discusses how activities moved out from homes in recent years to public and commercial spaces and how Covid-19 found many homes inadequate for multiple functions. The chapter reviews what could constitute a sustainable home or what home could support a sustainable lifestyle. It presents a perspective as seen from Israel.

Chapter 5 focuses on communities and neighborhoods. It opens with describing communities in Chinese cities, which are explicitly related to residential districts with clear physical boundaries, guarded entrances, community-based governmental and non-governmental organizations, and basic public facilities like stores and parks. The Chinese household registration system requires communities to play a major role in providing services, like childcare, schooling, and healthcare. They also enable the implementation of urban, local policies through a high degree of concerted collective awareness and action, which provided essential support during the pandemic for safety, information, food provisioning, healthcare, public space, and facilities. Urban communities acquired a new role in emergency response, risk management, and city resilience during the lockdown period, a trend highly likely to persist. Chinese cities are governed via these community grids, in which communities function as micro-drivers behind urban transitions.

Chapter 6 focuses on cities and on how the accelerating and decelerating trends during the pandemic are changing how cities are functioning. It pays particular attention to the implications of digitalization and decentralization on the future of city centers, which lost businesses and functions. It reviews approaches to sustainable cities and discusses whether cities will be more or less sustainable if accelerated digitalization increases home working, online shopping, and entertainment. It suggests that city centers may need to redefine their roles and that neighborhood will take the lead, following preference for local activities and micro-mobility. It presents a perspective as seen from Israel.

Chapter 7 makes the connection between changing lifestyles and another global threat, climate change. Although the climate and Covid-19 crises are very different in time scale and in their effects, there are also clear connections. The first is that both crises are, to a certain extent, the consequence of our present lifestyles. Dominant climate policies have primarily focused on technological solutions like renewable energy, electric vehicles, and home insulation. Yet different stakeholders increasingly recognize that, in addition to technologies, lifestyle changes are necessary, however hard they are to achieve. The second connection is that the lifestyle effects of the Covid-19 crisis have effects on GHG emissions. Government interventions brought sustainability benefits, while opening a rare opportunity for encouraging carbon-neutral lifestyle changes. However, the pandemic also forced negative sustainable behaviors, such as the resurgence of individual mobility and a throwaway culture of disposable, plastics. It presents an international perspective.

Chapter 8 focuses on responses to the challenges and opportunities for systemic transitions toward sustainability through policy measures. China was the first country that coped with the virus outbreak through countrywide lockdown measures and returned to normal social and economic activities. Economic recovery was its priority, but it faced a dramatically changed world environment. The chapter discusses how this opportunity could be harnessed to promote green growth, high-quality development, sustainable lifestyles, and low-carbon transitions. It presents major policy instruments deployed in China and internationally at the local and national levels, which could contribute to a more sustainable future.

Chapter 9 addresses three major questions. How will emerging trends contribute to sustainable lifestyles? Second, how can governments and policy makers facilitate and modify institutional structures and infrastructures to harness and promote emerging sustainable lifestyles? Third, is Covid-19-embedded policymaking going in the right direction? For each of these questions, sustainability encompasses environmental, social, economic, ethical, and wellbeing aspects.

References

Banerjee, D., and Rai, M., 2020. Social isolation in Covid-19: The impact of loneliness. *International Journal of Social Psychiatry*, 66, 525–527. https://doi.org/10.1177/0020764020922269.

Boons, F., Burgess, M., Ehgartner, U., Hirth, S., Hodson, M., and Holmes, H., 2020. *Covid-19, Changing Social Practices and the Transition to Sustainable Production and Consumption*. Version 1.0. Manchester: Sustainable Consumption Institute.

Bradbury-Jones, C., and Isham, L., 2020. The pandemic paradox: The consequences of COVID-19 on domestic violence. *Journal of Clinical Nursing*, 29, 2047–2049. https://doi.org/10.1111/jocn.15296.

Brooks, S. K., Webster, R. K., Smith, L. E., Woodland, L., Wessely, S., Greenberg, N., and Rubin, G. J., 2020. The psychological impact of quarantine and how to reduce it: Rapid review of the evidence. *The Lancet*, 395 No. 10227, 912–920. https://doi.org/10.1016/S0140-6736(20)30460-8.

CNNIC, 2020. Statistical report on internet development in China. http://cnnic.com.cn/IDR/ReportDownloads/201911/P020191112538996067898.pdf (Accessed 08.09.20).

Cohen, M., 2020. Does the COVID-19 outbreak mark the onset of a sustainable consumption transition? *Sustainability: Science, Practice and Policy*, 16 No. 1, 1–3. https://doi.org/10.1080/15487733.2020.1740472.

The Economist, 2020. The great reversal. Covid-19 is undoing years of progress in curbing global poverty. https://www.economist.com/international/2020/05/23/covid-19-is-undoing-years-of-progress-in-curbing-global-poverty (Accessed 23.05.20).

Giddens, A., 1991. *Modernity and Self-Identity*. Cambridge: Polity Press.

Huang, Y., and Zhao, N., 2020. Generalized anxiety disorder, depressive symptoms and sleep quality during COVID-19 outbreak in China: A web-based cross-sectional survey. *Psychiatry Research*, 288, 112954. https://doi.org/10.1016/j.psychres.2020.112954.

IBGE, 2019. Pesquisa Nacional por Amostra de Domicílios (PNAD) Contínua: Características dos Domicílios e dos Moradores. https://biblioteca.ibge.gov.br/visualizacao/livros/liv101707_informativo.pdf (Accessed 09.06.20).

IMF, 2020. Real GDP growth. WEO October 2020. www.imf.org/external/datamapper/NGDP_RPCH@WEO/OEMDC/ADVEC/WEOWORLD (Accessed 12.03.20).

Kennett-Hensel, P. A., Sneath, J. Z., and Lacey, R., 2012. Liminality and consumption in the aftermath of a natural disaster. *Journal of Consumer Marketing*, 29 No. 1, 52–63. https://doi.org/10.1108/07363761211193046.

LeQuéré, C., Jackson, R. B., Jones, M. W., Smith, A. J. P., Abernethy, S., Andrew, R. M. et al., 2020. Temporary reduction in daily global CO_2 emissions during the COVID-19 forced confinement. *Nature Climate Change*, 10, 647–653. https://doi.org/10.1038/s41558-020-0797-x.

Markard, J., and Rosenbloom, D., 2020. A tale of two crises: COVID-19 and climate. *Sustainability: Science, Practice and Policy*, 16 No. 1, 53–60. https://doi.org/10.1080/15487733.2020.1765679.

Nakada, L. Y. K., and Urban, R. C., 2020. COVID-19 pandemic: Impacts on the air quality during the partial lockdown in São Paulo state, Brazil. *Science of the Total Environment*, 730. https://doi.org/10.1016/j.scitotenv.2020.139087.

Neves, L. M. P., 2004. Cleanness, pollution and disgust in modern industrial societies: The Brazilian case. *Journal of Consumer Culture*, 4 No. 3, 385–405. https://doi.org/10.1177/1469540504046525.

Power, K., 2020. The COVID-19 pandemic has increased the care burden of women and families. *Sustainability: Science, Practice and Policy*, 16 No. 1, 67–73. https://doi.org/10.1080/15487733.2020.1776561.

Progrebna, G., and Kharlamov, A., 2020. The impact of cross-cultural differences in handwashing patterns on the COVID-19 outbreak magnitude. https://doi.org/10.13140/RG.2.2.23764.96649.

SSPP (Sustainability: Science, Practice and Policy). (2020) Special Issue: COVID-19 as a Catalyst for a Sustainability Transition.

Torales, J., O'Higgins, M., Castaldelli-Maia, J., and Ventriglio, A., 2020. The outbreak of COVID-19 coronavirus and its impact on global mental health. *International Journal of Social Psychiatry*, 66, 317–320. https://doi.org/10.1177/0020764020915212.

WEF (World Economic Forum)-Ipsos survey, 2020. Nearly 9 in 10 people globally want a more sustainable world post-Covid-19. https://www.weforum.org/press/2020/09/nearly-9-in-10-people-globally-want-a-more-sustainable-and-equitable-world-post-covid-19/ (Accessed 09.06.20).

WWF (World Wildlife Fund)-GlobeScan survey, 2020. Study finds people want to make healthy and sustainable living choices but do not know where to start. https://globescan.com/people-want-healthy-sustainable-living-choices-2020/ (Accessed 10.08.2020).

2   Post-Covid-19 trends

Acceleration, deceleration, and innovation

No relevant territory or country of the global map was left untouched by the pandemic. From the homogeneous Global North, including Australia and New Zealand, to the rapidly modernizing Global East world, to the vastly heterogeneous Global South, all nations underwent similar processes and patterns of contagion. Similarly, across the globe, societies adopted roughly the same standard protocols and measures, and were reshaped by the same forces and tendencies that redefined priorities, social norms, and broader attitudes and behaviors toward aspects of lifestyles. This chapter summarizes major trends that took place in these different contexts while echoing what occurred in the "developed" Northern Hemisphere. It portrays accelerated trends, decelerated tendencies, and innovative, disruptive phenomena in various regions, including the peculiar flavor of those countries we focus on. Through the interconnections by social media and other interfaces, these trends transcended national boundaries.

Accelerating, decelerating, and innovative trends

Multiple spheres of life were affected by the pandemic. Some of the effects on daily life were generated by the emergence of new social characteristics and attitudes during the pandemic and are likely to continue after health risks no longer pose a threat to safety. Others relate to relationships between government and civil society within changing social and political frameworks. This chapter reviews how Covid-19 accelerated some trends, decelerated others, and generated some new ones, and the effects of these processes on ten domains of daily practices. We are especially interested in those that are likely to continue after countries exit lockdowns and that will lead to different lifestyles currently taking root. Our review of these changes focuses on whether they may promote or restrain more sustainable lifestyles (developed further in Chapter 7).

Accelerated trends

Accelerated trends refer to developments that were already in progress before Covid-19 and that scaled up and intensified during the pandemic. Some of the accelerated trends were already mainstream, affecting a significant part of the population. Others underwent very significant acceleration but remain peripheral, affecting a limited number of people.

The chapter focuses on two major accelerated trends: digitalization and mindful consumption, both of which cut across multiple domains of practice and are likely to affect attitudes, beliefs, and behaviors.

Digitalization

Digitalization of functional and emotional spheres of life was already happening before Covid-19 hit humanity. It was transforming work, leisure, shopping, and intimate relationships through remote forms of engagement.

Multiple everyday life practices were already operating in internet-embedded environments in society. They not only provided functional and instrumental purposes but also influenced perceptions and attitudes, and provided an expression of identity and thus played a communicative role about the individual. Digitalization is not just about the adoption of technological tools but provides opportunities to express self-representation (and fulfill the need for self-esteem), a sense of belonging to a virtual or real community, and self-actualization through instilling a sense of participation and accomplishment. Digitalization enables communications on multiple platforms, generating interactions and bringing together participants with common shared social meanings. Some scholars have conceptualized it not as a feature of lifestyles but as a lifestyle in itself, considering "digital nomads" to be its quintessential character (Thompson, 2019).

The digitalization of domains of practice has converged with, and amplified, parallel trends in specific lifestyle expectations. Telework, for example, corresponds with increasing demands for autonomy in the workplace and transference of responsibility to workers to achieve higher productivity, motivation, and employee satisfaction. Remote work requires achieving results or goals rather than meeting schedules or following protocols. This new balance of employer-employee relationships undermines hierarchy and the concentration of responsibilities and liabilities in the workplace, and places greater emphasis on self-realization and self-actualization.

Telework was easier to implement in Europe and North America given the degree of quality internet penetration, but it was in the Global South where it showed the fastest growing rates. For example, early incidence accounts of telework in Brazil indicated that less than two in every eight would be able to adopt it (Mercer Consulting, 2020), yet by the end of the year projections indicate a 30% growth in work from home (Miceli, 2020). Moreover, official reports consider that nearly 23% of occupations, involving 21 million economically active Brazilians, are highly likely to integrate telework once the pandemic is over (IPEA, 2020). A digital home-office may provide greater comfort, convenience, and a better use of time. Saving the time and energy otherwise spent in daily commuting through traffic congestion contributes to personal wellbeing and offers the opportunity for the employee to spend more time with family and to participate in family routines, which contributes to a better work-life balance.

The migration of shopping and food provision to online channels was an outstanding feature of the pandemic – a process that intensified around the world and particularly in the Global South and Global East (McKinsey, 2020). In Brazil, e-commerce transactions during the first 6 months grew by 40% compared to the same period of 2019 (Ebit/Nielsen, 2020), with one-third of online shoppers being first-timers (Comscore, 2020). The reduction of shopping as an activity and its transfer to purchasing online have the potential for better planned, less impulse-driven purchasing. Conscious consumerism based on rational planning enables a distinction to be made between essential and non-essential items, hopefully encouraging less purchasing through a conspicuous attitude to acquiring goods, more cost conscious, and less "throwaway". If this becomes a general change of attitude in food and home provisioning activities, it could encourage more sustainable patterns of consumption.

Online gaming has been growing rapidly over the past decades. However, gaming during Covid-19 helped to meet the need for social belonging and connectivity. Video-messaging, virtual parties, and in-group movie watching provided opportunities for virtual socialization and social connectivity, particularly among younger age groups. Online dating apps such as Tinder attracted romance seeking, which accelerated in their evolution and magnitude during the pandemic.

Digital learning was already well developed prior to Covid-19, but during the pandemic, there was a boom in e-learning. Experience with tele-education for schooling during the pandemic met with mixed results, but distance learning for adults during free time under stay-at-home regimes increased significantly. Between 15% and 30% of the adult population in Brazil engaged in e-learning (McKinsey, 2020), but the use of remote teaching platforms only increased by a net 10 percentage points (Ipsos Global Trends, 2020).

More intensive and socially amplified digital mediation of many daily practices offers several benefits for sustainable lifestyles. The negative environmental effects of automobile-based mobility related to work, study, shopping, and leisure diminish. Increased productivity associated with tele-work may reduce resistance to a shorter workweek, enabling a redistribution of employment and worktime. A shorter working week can in turn contribute to strengthening relationships with family, community, and friends and to less materialistic, consumer-oriented forms of leisure and entertainment. Opinion surveys in India, China, and Latin America indicate a majoritarian appreciation of the opportunity to get closer to family and friends during the Covid-19-related restrictions on movement (Ipsos Global Trends, 2020).

Accelerated digitalization, however, has its downside and may generate social problems at the levels of the individual and society. At the individual level, telework can lead to feelings of isolation, alienation, and a propensity to feel overburdened, which could reduce motivation and wellbeing (Alfageme, 2020).

Over-reliance on virtual connectivity can result in the loss of social skills, lack of empathy, and the reinforcement of inner-group beliefs among heavy social media users. Social intolerance, poorer language and communicational skills, and the lack of social empathy do not constitute progress toward sustainability. Excessive internet or smartphone use raises concerns, particularly among children and adolescents, about technology-related addictive behaviors leading to problems of cognitive absorption, anxiety and depression, and disorders in sleep and physical health (Garcia-Priego et al., 2020).

At a societal level, rapid migration to a digital lifestyle accentuates social inequalities. In some societies, only a minority of economically active population holds jobs amenable to home-office. Among developing societies of Latin America, the vast majority are self-employed, informal workers, or employed in underfunded small firms, which excludes the possibility of teleworking (nearly 80% of total activities cannot be conducted from home; ECLAC, 2020). The same is true for tele-education. During quarantine, over 40 million households remained unconnected to the internet in Latin America, leaving 46% of children aged 5–12 with no means for tele-education (ECLAC, 2020). Inequalities are heightened by uneven computer literacy across classes, age cohorts, and professions. The extra expense for LAN house access to online services adds more financial pressure on low-class households' incomes, which were already under strain from the economic shutdown. Eight out of ten public school teachers in Brazil admitted to a lack of training and experience to adopt e-learning (Cafardo, 2020).

The boom of Internet and Communication Technologies (ICT) industries sparked by digitalization also has important environmental downturns, through a continued increase in the sales of computer, gadgets, webcams, and other equipment. Higher energy consumption for usage is obvious, but less explicit is the reinforcement of a business model based on a shorter lifetime of end-user equipment, contributing to the depletion of natural resources and the increase in e-waste.

The key to the acceleration of digitalization was the willingness of the public to adopt digital technologies and integrate them into current lifestyles. Government, the market, and civil society organizations played equally important roles. Government regulations permitted telework, distance learning, and operated programs for raising computer literacy within public schools. The market facilitated consumer digital engagement and provided infrastructure to harness the potential for mainstreaming digitalization. Business investment in widening internet coverage and broadband, together with decreasing prices for technology, lessened barriers to internet access and usage. Lastly, social institutions have played an important role in promoting digital inclusion, such as CDI/Recode, which started in Brazil a quarter century ago and has expanded to seven developing countries, reaching close to 2 million lower income individuals.

Mindful consumption

The main impacts of the pandemic on consumption, as mentioned earlier, were restraints on purchasing other than for food and essential items and the impact of digitalization on restraining conspicuous consumption and promoting conscious consumption. However, there has also been a very significant rise in mindful consumption as expressed in the preference for goods and services that give due consideration to social, environmental, ethical, and health aspects of consumer choice (Reid, 2020). Such trends already existed on a small scale where choices factored in effects on the environment, society, and personal health (Kronthal-Sacco and Whelan, 2019; SustainAbility Trends, 2020). During containment, the public expressed increased engagement in more mindful consumption, predominantly related to food and diet not only for their own health and protection but also for collective interests.

Consumers have consistently favored brands with corporate social responsibility (CSR) credentials, particularly in non-OECD countries (GlobeScan, 2020). If changing consumers' values and orientations help to understand this trend (Giddens, 1991; Beck, 2002), the critical leverage was provided by the corporate world that sympathized with CSR principles (Scherer and Palazzo, 2011) and by social movements that promoted the connection between private consumption and public benefits (Dubuisson-Quellier, 2013; Echegaray, 2016).

Recent surveys show that sustainability was already becoming accepted as a mainstream consideration. Nearly half of consumers in mature and developing societies buy (and continue buying) goods associated with sustainability (Accenture, 2020), even though the meaning of sustainability may differ between consumers (Grunert et al., 2014; Hanss and Böhm, 2012). The use of cognitive shortcuts (e.g., ecolabelling, standards formalization, certifications) has not always contributed to comprehension and choice-making (Horne, 2009). The predominant properties understood by consumers, and the main qualities communicated by labels, refer to the environmental assets of products. Covid-19 added ethical and health-related properties, accelerating gradual but constant growth of ethical purchasing, which in some mature markets like the UK had already experienced a 2.3 times growth rate over the last decade (Coop, 2019).

Mindful consumption goes beyond greening criteria, which largely characterize pro-environmental or pro-sustainability behavior. The pandemic did accentuate concern for the environment in consumer choice, particularly for food and household items, because the Covid-19 outbreak was connected with a failure between food supply and wildlife. However, the virus primarily triggered health-centered and immunity personal concerns. Dietary habits, product security (particularly for food), and the origin of production became major consumer concerns (see Mao et al., 2020).

Natural and organic food purchases

Diet change and immunity strengthening in response to Covid-19 reinforced ongoing trends of interest in healthy foods, mainly organic or natural. A total of 81% of shoppers in Brazil expressed interest in this trend (FMCG Gurus, 2020), and global market research has shown a continuous rise of interest in organic food and willingness to pay more for sustainable ingredients since 2016. Major growth beyond 2020 was anticipated as younger generations showed strong interest in these trends (Ipsos Global Trends, 2020).

Diet reform usually includes removal of meat from the diet, considered as less healthy and eco-friendly. Businesses offered alternative meat and dairy products in the market, and social movements including vegetarians, vegans, and environmental groups promoted and facilitated consumer access to them. The reduction of meat and dairy products in the diet had already become accepted as a mainstream concern, partly related to understanding the lower footprint of vegetal meats, partly because celebrities raised public awareness about animal welfare, and partly because the availability of plant-based meats rapidly adopted by fast-food chains. The pandemic intensified the translation of diet change-consciousness into a migration toward plant-based foods, particularly in mature markets like the US, with plant-based meat and plant-based dairy sales rising to 280% and 477%, respectively (Nielsen, 2020). Producer sources estimate that 23% more people shifted to eating plant-based foods in response to Covid-19 in the US (Danziger, 2020); a similar proportion did so in Europe (+25%; cf. Allen and Warwick, 2020). Projections for Global South markets like Argentina, Brazil, or Colombia are equally optimistic, with expected growth rates in vegetal protein in the range of 65%–78%, respectively (Spencer, 2020).

Food-centered mindful consumption is not new. In Brazil, the Slow Food movement, agro-ecological networks like Kairós, solidarity purchase clubs, and children's rights NGOs like Alana have raised awareness, raised the costs of misleading advertising, lobbied intensively for proper labelling of products at retail chains, and connected small-scale producers with consumers. Government policies that have favored family agriculture since the mid-2000s required that 70% of federal school-lunch funds be spent on unprocessed food and that at least 30% must come from family farms (Huber, 2016). During the pandemic, the government maintained the distribution of lunch boxes following these criteria.

There are a number of barriers to mindful consumption. Prices of organic, ethical, and sustainable products are often at a premium, which is not affordable at a time of rising unemployment and cuts in spending. In the face of an adverse financial situation, concerns about contagiousness, and the wish to reduce the frequency of food provisioning trips to supermarkets, consumers gave priority to purchasing foodstuffs with long expiration dates such as industrialized, frozen processed foods (+21%–23% sales growth in Brazil; cf. Nielsen, 2020). Concomitantly, natural, plant-based cleaning products suffered a loss of customers since they were perceived as less safe or less efficient for disinfection as compared to conventional, petrochemical-based products (Nielsen, 2020).

Localism

A growing preference for local products and shorter supply chains was already developing in consumer markets in Brazil as well as in many other Latin American countries before containment was imposed during the pandemic. Community consumption, as it was dubbed, reflected a concern about the origin of products as much as an ethical standing in defense of an expanded notion of community (Szmigin et al., 2003). During the virus outbreak, localization of consumption converged with concerns about product security and trust in globalized, long supply chains. Online conversations in Twitter about shopping locally rose by 440%, for example (Brandwatch, 2020). Public opinion surveys revealed in early 2020 that 70% of the population stated a preference for locally grown food (Chinese consumers exceeding other nationalities in their preference for community consumption; Ipsos Global Trends, 2020). Studies in the US showed a 53% rise in people more likely to buy from local stores than from national retailers during the pandemic (Zypmedia, 2020).

Localization trend was further reinforced by the embracement of home-cooking during the pandemic. This new behavior also leveraged different, mindful habits such as the reuse of leftover food and the reduction of food waste. Surveys showed that 47% of adults in the US shifted to eating more home-made food (IFIC/FoodInsight, 2020). In Brazil, a third of the population cooked for the first time during the quarantine (Nielsen, 2020).

Personal savings

Saving money has lost favor in recent years, especially among young people who earned well and spent their income on enjoying life. Covid-19 exposed the precariousness of this attitude and induced changes. Nearly one in three Americans who received the stimulus check put it in savings (Fitzgerald, 2020) and two in five Americans acknowledge they are now saving more than before, exceeding the percentages who are spending or investing more (El Issa, 2020). People in Britain reacted similarly, with one-third saving money during the pandemic and another one-fifth planning to do so (Ipsos-Mori, 2020). In Brazil, savings share of total GDP reached record levels amid the pandemic (15.5%), the highest volume in 5 years (Pamplona and Cucolo, 2020). On the other hand, between 25% and 40% of Europeans responded that they do not have any savings (ING, 2020).

Decelerated trends

Decelerated trends include the reduction of the growing popularity of dense cities and the separation of in-home from out-of-home activities. These two trends will be treated in more detail in Chapters 4 and 6 on Home and Cities. In this section, we focus on two other decelerated trends, both with substantial effects on lifestyles and behavior: setbacks in the ongoing processes toward alleviation of poverty and reducing inequalities in wealth and education (which we call "social marginalization"); and the setbacks in the evolution of "alternative economies", in particular, the sharing economy and the circular economy.

Social marginalization

Before Covid-19, the world was progressing well on some of the Sustainable Development Goals (SDGs). There were encouraging signs that fewer people were suffering from poverty and hunger. However, the virus outbreak brought a major setback in achieving SDGs 1, 2, 5, and 10 because of the impacts of containment measures that led to rising hunger and deeper social and gender inequality. Drainage in family resources due to layoffs, cuts in salaries, and the closing down of small business, together with lockdown measures that inhibited the works of the informal economy, reverted any previous progress. Covid-19 also affected other SDG like number 4 (quality education) when schools were closed, aggravating disparities for the student population without access to e-learning alternatives,

Sliding back into poverty as a result of the pandemic may affect millions around the world. Predictions for Latin America show that between 29 million and 45 million people will drop out of the middle classes and fall back into poverty (ECLAC, 2020; UN News, 2020) with all the associated misery and social inequality (Fisher and Bubola, 2020; van Dorn et al., 2020). These disparities were accentuated where people could not telework, particularly the large number of informal workers in the Global South, who lost their sources of income and experienced higher health risks. People living in shantytowns and overpopulated tenements that lack sewage and drinking water could not implement social isolation and hygiene policies. Fifteen percent of households (31 million) in Brazil lack domestic potable water; furthermore, over 6% of the total population (11.5 million) live in overcrowded households (average three persons per room), which rises to 13% of the population in the North/Northeast regions (IBGE, 2019). Predictably, these regions were the worst affected by the virus outbreak. Constraint living conditions intensified the unequal distribution of risks in society and resulted in a radical social bias of Covid-19 impacts that penalized the poor.

Gender equality also experienced a regression during the pandemic (Alon et al., 2020; Madgavkar et al., 2020; Power, 2020) due to overburdening female caring activities with home-schooling and home cleaning, in addition to higher female exposure to furloughs and unemployment. Women in India spent 30% more time on family responsibilities and women in the US an additional 2 hours (Madgavkar et al., 2020). In Brazil, the jobless difference across gender was above 7 percentage points against females (IBGE, 2020). Although particularly acute in the Global South, gender inequalities were heightened by the pandemic across all societies.

Regressions in sharing and circular economies

Many sustainability scholars welcomed advances in collaborative consumption and sharing through online platforms as a positive development toward reducing the volume of material goods (Belk, 2014). The boom in internet-mediation among owners and users of assets to enable easy and inexpensive access to underutilized goods offered an example of how multiple agents in mainstream economic systems could overcome locked-in situations. The sharing economy enabled a transformation of actors, facilitators, and infrastructure toward more sustainable outcomes (Hamari et al., 2016). Although primarily for commercial purposes, collaborative consumption had some non-commercial purposes, such as strengthening community relations, sharing non-monetary and non-materialistic activities, and practices of giving, bartering, swapping, and trading goods and services (Belk, 2014).

Similarly, sustainability scholars welcomed advances in the circular economy. The circular economy seeks to optimize the life cycle of goods by closing loops, extending product life, augmenting resource efficiency, reducing waste, and dematerializing the economy. The sharing economy highlights the role of access to goods over ownership; the circular economy underscores the role of production and models for utilization through long-lasting design, maintenance, repair, reuse, remanufacturing, refurbishing, and recycling (Geissdoerfer et al., 2017). Engagement of the consumer in the circular economy involves repairing, reusing, preferring recycled goods, and cooperating with waste recycling.

The pandemic undermined many of these behaviors. Concerns about hygiene and increased takeaway and food delivery hugely increased the use of disposable packaging and plastic materials (The Economist, 2020). Food is packaged and delivered in non-compostable, hard-to-recycle single-use containers, including disposable plastic utensils.

Social isolation policies that put a premium on health preservation and avoidance of contagion challenged the very essence of the sharing and circular models: the relevance and ubiquity of access and utilization. Coach surfing and mediated peer-to-peer renting of accommodation came to a halt. Home-office schemes and the closing of schools and workplaces removed the demand for carpooling and shared rides. Time banks and tools swapping or renting became non-operational because of circulation restrictions. Co-working stations emptied and went bankrupt as the freelancing economy adapted to work-from-home. The pandemic has damaged numerous sharing economy businesses, particularly the industry's hallmarks (Conger and Griffith, 2020). Re-use or sharing of previously used goods clashes with physical distancing and minimizing exposure to infection through contact with unknown surfaces. Similar hurdles impair the progress of repairing and recycling practices. Following unions' requests, solid waste sorting, recollection, and recycling centers were suspended due to contagion concerns, which led to the interruption in home waste segregation and broader recycling practices. This occurred in different contexts, be it Brazil, US, or Italy.

For those feeling "consumption revenge" to compensate for months of delayed, repressed wants due to the closure of shopping malls and the prioritization of essential items, the availability of used goods and services from the sharing or circular economies would not be a gratifying alternative. Policies for containment and isolation thereby reinforced non-sustainable behavior patterns, such as individualized mobility, preference for product ownership over access, and a throwaway mentality. Covid-19 has facilitated the activation of negative attitudes, raised the costs of accessibility to sharing, repairing, reusing, or recycling options, and frozen mechanisms and available infrastructure of circular and collaborative economies. Single occupancy car use came back, purchase of tools substituted lending or borrowing, and irresponsible disposal behavior became the accepted social norm as individual, single-use plastic receptacles and utensils become the rule again (The Economist, 2020).

Innovative, unexpected trends

The pandemic gave rise to disruptions and radical changes that no previous theory, data, or market analysis foresaw. We identify four of those that are relevant to sustainable lifestyles: the health-centric reconceptualization of wellbeing, the normalization of social isolation, the blurring of boundaries defining what is home, and the (transient?) regression toward a passive civic culture, which accepts increased control authority by government.

Reconceptualizing wellbeing

Emphasis on hygiene and health transformed the concept of wellbeing into a germ/virus-free, protected environment, free of contamination and restricted to known and trusted persons. This modification in cultural codes and meanings was unforeseen by scholars or marketing practitioners concerned with consumer or societal trends.

News headlines and academic debate have usually focused on the tension and dilemma brought by quarantine measures between protection of public health and protection of economic health. Yet there is another point of tension brought about during the pandemic: between the protection from the risk of infection and protection of mental and physical health. Longer quarantines raised levels of anxiety, depression, overweight and obesity, sleep disorders, alcoholism, and addictions (Fernández et al., 2020; Torrente et al., 2020).

Normalization of social isolation

Responses by governments to Covid-19 required people to live physically distanced and socially isolated, limited to a small and intimate circle of trusted (clean-reliable) contacts. The effects upon lifestyles, social connections, and social life were enormous. Even when partially replaced through internet-mediated means like video chats, online family gatherings, and Zoom parties, the benefits and quality of socializing were reduced. Migrating interpersonal relationships online impoverished the quality and the quantity of communication as much as its socio-psychological benefits (Waytz and Gray, 2018). A minority of single-residents used alternative coping behaviors by sheltering with parents or friends, dubbed "quaranteaming".

Although reduced travel for social connectivity could be seen as good news for the environment, it has had adverse consequences for the subjective wellbeing of individuals. Loneliness was already recognized as a societal problem by leading to a loss of social abilities, heightened sense of vulnerability, particularly among the youngest and oldest cohorts (Torales et al., 2020), and aggravation of mental and physical health.

Homebody life

The term "homebody" is usually associated with a personal preference for remaining at home and an aversion to adventure (Cambridge dictionary). Covid-19 turned everybody into homebodies. Boundaries between household, workplace, shopping, studying, fitness room, and leisure area became blurred and all activities took place within the home. Lockdown policies placed the home at the center of everyone's lives, becoming the single ambience hosting multiple functions. Homebody practices were imposed on individuals and families forced to stay 24 hours a day, 7 days a week in spaces that had to accommodate multiple functions of sleeping, relaxing, dressing, showering, and many more (see further Chapter 4). Forced homebody life resulted in very different perceptions and experiences, depending on the number of residents within the household and the space available. The concentration of multiple activities into the home shaped behavioral patterns and affected values and beliefs. Variations related to households' member size and complexity in having to perform multiple roles, responsibilities, and life goals in a single, ill-equipped space (Prime et al., 2020).

During the pandemic itself, necessary home-centric requirements provided a sense of security and control while reducing the feelings of time-pressure typical of routines in pre-Covid-19 times. People recognized the benefits of slowing down schedules and activities, and the availability of free time saved from commuting, which became immediate payoffs (as reflected by opinion surveys). They enabled households to reflect on prior dynamics and reconsider the balance of life in family households. Responses to home confinement pointed to the value of having free time, the ability to achieve a work-life balance, the opportunities for fruitful social interaction with spouses, home-mates, and children, time to resume hobbies and pastimes that nurture time beyond work or study, and taking small steps toward increasing personal wellness. The stay-at-home recommendations removed barriers to engage in low budget or no budget activities and provided opportunities for non-for-profit time-use and non-commercialized wellbeing and sociability that may continue and ultimately reduce ecological footprints.

Forced homebody life, however, also revealed the limitations of transferring multiple roles, responsibilities, and life goals into a single, ill-equipped space. Managing simultaneous activities within limited space put both task performance efficacy and interpersonal conviviality to the test. Overcrowded households typical of poor communities were subject to higher contagion, poorer performance, and problematic domestic governance issues (Cimini and Botts, 2020). Adapting household structures remains unfeasible for many and requires additional spending when family budgets may be reduced by the economic shutdown. Tensions within households rose and, not surprisingly, lockdowns were accompanied by a rise in domestic violence, abuse of children, divorces, and other expressions of wellbeing deterioration (Bradbury-Jones and Isham, 2020).

Resurgence of elite-directed, passive citizenship

The public health emergency and the enforcement of social distancing regenerated the limited and conservative notions of citizenship of past times, opening the door to unexpected changes in the meaning and nature of citizen engagement. In pre-Covid-19 times, critical citizenship gained muscle (Norris, 1999), combining a high adherence to principles of democratic living, a personalization of politics based on engagements related to lifestyles matters (Bennet, 2012), and low trust in authorities. This model of civic behavior was premised on individualization theory (Beck, 2002) that acknowledged the impotence or lack of interest by governments to address personal important issues of daily life, not abstract ideological themes but concrete problems such as climate change, gender inequality, transgender issues, overweight, pandemic, or genetically modified organisms (GMOs) and food safety. Perceived as short of resources, understanding, or will to solve this type of problems, citizens positioned themselves as elite-challenging rather than elite-oriented publics (Inglehart, 1990). Consequently, individuals bypassed the state or government as interlocutor for producing relevant outcomes in non-electoral times, and switched their target for activism to other agents like NGOs and corporations, while attributing themselves with self-responsibility for addressing matters directly, at the market shelves, corporate headquarters, or social media.

The pandemic blueprint to fight the contagion outbreak placed the state back in at the center stage. The return of formal political authority became obvious through the coordination of multiple measures of national or local enforcement, from stating the triple shock (referred to in Chapter 1) to distributing emergency relief funds and tax exemptions to offset the economic hardship across segments, to designing and applying bailout schemes and – eventually – green new plans as post-Covid-19 recovery policy. Street policing and prosecution of quarantine breakers, government set-up of solidarity networks to help vulnerable publics, and digital and physical surveillance of individuals, families, and commercial establishments to meet lockdown protocols stand out among the steps that brought the state back in, usually with strong support of society. Public trust in government, political establishment and representative institutions, and the state resurged, evoking wartime "rally-around-the-flag" effects.

Critical citizenship left room to the comeback of something unexpected: a resurgence of deferential, social control-accepting political culture in many Global South and Global East contexts that resembles descriptions of conventional "civic culture" (Almond and Verba, 1963). Characterized by law-obeying and social order preferences, political neutrality, elite-directed acquiescence, and high institutional confidence, the pandemic's reshuffle of political subjectivity froze or slowed down not just activism but also individuals' political autonomy and support for democracy once the reduction of civil liberties and political rights is accepted as normal. This has consequences for sustainability in general.

On the one hand, a demobilized citizenry around hot causes like climate change, environmental protection, and biodiversity loss (that was gaining momentum back again just before the virus outbreak) may regress society advances and commitments by governments and the market. More tangibly, governments harnessed this situation to relax environmental protection rules (Knickmeyer et al., 2020). In Brazil, lockdown in the public sector left forests and indigenous communities unprotected as environmental regulation authorities remained under quarantine, which resulted in a rise of deforestation (Fonseca et al., 2020) and a disproportional number of indigenous people infected by the virus (Instituto SocioAmbiental, 2020). Likewise, tourism-drain in Africa resulted in underfunding of wildlife anti-poaching and anti-deforestation and rural settlements threatening biodiversity in the short and long terms.

On the other hand, passive acceptance by the public of increased surveillance and social control over individuals in order to fight the virus is easing the trend toward autocratic regimes across many countries. Several Eastern European governments imposed illiberal models of electoral authoritarianism that cancels rights and freedoms. Argentina weakened check-and-balance institutions by maintaining Congress or the Supreme Court partially closed for the sake of social isolation measures. In the Global South, support for militarized ghetto-type policies toward slums as outbreak control measures crystallized social inequities. Moreover, the revival of a strong state rhetoric leaves no tolerance for denunciations of procurement corruption, which leaves social projects further underfunded.

Table 2.1 summarizes accelerating, decelerating, and novel trends during the pandemic and their impacts on ten domains of practice.

Table 2.1 Trends acceleration, deceleration, and innovation across main domains of practice

	Acceleration	Deceleration	Innovation
Work	Digitalization through home-office.	Sharing economy options like co-working stations.	Work-life balance. Potential reduction of workweek.
Family/social connections	Digitalization through social media and internet-mediated channels. Demand for slower pace, simplified life addressed through stay-at-home. Smaller families. Responsible parenthood.	Enthroning of privacy and autonomy. Decreases in domestic violence. Gains toward gender equality halted due to higher shouldering of caring burden.	Social cocooning as social norm. Mindful consumption channelized through non-commodity-based, dematerialized relationships with family members. Rise in community and neighborly attachments.
Wellness	Digitalization through virtual fitness and wellness classes. Touchless approach expanded as social convention and as consumer demand. Overweight and obesity increases.	Social learning based on empathy, sharing, emotional development. Controls over internet addictions. Active lifestyles thereby triggering sedentarism and mental/physical problems.	Rise in sufficiency awareness. Renewed appreciation of savings over spending. Non-monetary-based mental and physical wellness routines (hikes, family life, and meditation).
Leisure	Digitalization through immersive tech, home-entertainment, and online events.	Group, peer-based on-site live entertainment	Revalorization of outdoor, natural surroundings, local trips. Embracement of "staycations". Mindful consumption through revival of DIY hobbies & nostalgia-based board games.

(Continued)

Table 2.1 (Continued)

	Acceleration	Deceleration	Innovation
Education	Digitalization through remote schooling. Heightened inequality between public versus private schools students	Meeting quality educational targets as per SDG #4.	Mainstreaming of distance learning through digitalization.
Mobility	Automobile single-rider individualization.	Sharing economy options like carpooling and car-sharing. Public transportation system use due to contagion fears and bureaucracy for use.	Individualized micro-mobility through walking/biking, favored by overnight recycling of infrastructure as bikeways.
Healthcare	Digitalization through telemedicine, remote psychotherapy and mental health management. Social isolation illness like loneliness, depression, anxiety.	Defunding of public health system reverted. Broader notions of healthy lifestyles now reduced to germ prevention and hygiene protocols.	Rise of health prevention and attention within a universal basic services perspective. Media and business linkage of human mistreatment of wildlife areas with virus outbreaks that reinforces a more relevant, persuasive narrative in favor of environmental protection.
Food and home provision and consumption	Digitalization through e-commerce and food delivery. Mindful consumption through diet change, health-oriented purchases. Yet, anti-mindful consumption buying based on industrialized, processed food and high alcohol intake.	Solid waste reduction and segregation practices. Circular economy practices at the individual-level like reusing or repairing.	Implementation of universal basic income. Consumer production (home-cooking) and heavy localization of purchases. Mainstreaming of plant-based diets. Food waste better management.

Housing	Single-resident apartments	Densification and small-size living units freeze. Home as back-up stage for out-of-home life.	Home as center of multiple activities. Desertion of cities. Revaluation of rural, suburban areas, naturally aired/lightened habitats.
Citizenship	Autocratization of governance. Increased surveillance and social control. Political consumerism. Digitalization-led click activism through social media.	Elite-challenging citizens through forced demobilization.	Acquiescence to erosion of civil liberties and political rights. Return of state authority and welfare state policies. Rise in prosocial, self-initiated networked actions in solidarity to vulnerable groups and favoring community interest. Environmental policy flexibilization co-occurring with Green New Deal discussions.

References

Accenture, 2020. COVID-19 increasing consumers' focus on "ethical consumption," Accenture survey finds. https://newsroom.accenture.com/news/Covid-19-increasing-consumers-focus-on-ethical-consumption-accenture-survey-finds.htm (Accessed 09.01.20).

Alfageme, A., 2020. O Sonho do 'home office' Vira Pesadelo Na Pandemia. *El País.* https://brasil.elpais.com/sociedade/2020-08-09/o-teletrabalho-nao-era-isto.html (Accessed 28.09.20).

Allen, R., and Warwick, C., 2020. A plant-based future? *Kantar Talks 19.* https://consulting.kantar.com/wp-content/uploads/2019/10/Kantartalks_A-plant-based-future_2019.pdf (Accessed 28.07.20).

Almond, G., and Verba, S., 1963. *The Civic Culture: Political Attitudes and Democracy in Five Nations.* Princeton: Princeton University Press.

Alon, T. M., Doepke, M., Olmstead-Rumsey, J., and Tertilt, M., 2020. The impact of COVID-19 on gender equality No. w26947. *National Bureau of Economic Research.* https://doi.org/10.3386/w26947.

Beck, U., 2002. *Individualization: Institutionalized Individualism and Its Social and Political Consequences.* Thousand Oaks, CA: Sage, v. 13.

Belk, R., 2014. You are what you can access: Sharing and collaborative consumption online. *Journal of Business Research,* 67 No. 8, 1595–1600. https://doi.org/10.1016/j.jbusres.2013.10.001.

Bennett, W. L., 2012. The personalization of politics: Political identity, social media, and changing patterns of participation. *The Annals of the American Academy of Political and Social Science,* 644 No. 1, 20–39. https://doi.org/10.1177/0002716212451428.

Bradbury-Jones, C., and Isham, L., 2020. The pandemic paradox: The consequences of COVID-19 on domestic violence. *Journal of Clinical Nursing,* 29, 2047–2049. https://doi.org/10.1111/jocn.15296.

Brandwatch, 2020. Sustainable, local, ethical: How is Covid-19 changing our shopping habits? www.brandwatch.com/blog/react-Covid-19-sustainable-shopping/ (Accessed 09.03.20).

Cafardo, R., 2020. Oito Em Cada Dez Professores Não Se Sentem Preparados Para Ensinar Online. *O Estado de S.Paulo.* https://educacao.estadao.com.br/noticias/geral,oito-em-cada-dez-professores-nao-se-sentem-preparados-para-ensinar-online,700033050 (Accessed 20.05.20).

Cimini, K., and Botts, J., 2020. Close quarters: California's overcrowded homes fuel spread of coronavirus among workers. *CalMatters.* https://calmatters.org/projects/overcrowded-housing-california-coronavirus-essential-worker/ (Accessed 09.06.20).

Comscore, 2020. Consumo de Mídia Durante a Pandemia De Coronavirus No Brasil. www.comscore.com/por/Insights/Blog/Consumo-de-midia-durante-a-pandemia-de-coronavirus (Accessed 19.05.20).

Conger, K., and Griffith, E., 2020. The results are in for the sharing economy. They are ugly. *The New York Times.* www.nytimes.com/2020/05/07/technology/the-results-are-in-for-the-sharing-economy-they-are-ugly.html?searchResultPosition=2 (Accessed 09.04.20).

Coop, 2019. Ethical consumer markets report. www.ethicalconsumer.org/research-hub/uk-ethical-consumer-markets-report (Accessed 08.07.20).

Danziger, L., 2020. Survey: 23% of Americans eating more plant-based foods during COVID-19. *The Beet*, 6 May. https://thebeet.com/survey-23-percent-of-americans-eating-more-plant-based-foods-during-covid-19/ (Accessed 27.08.20).

Dubuisson-Quellier, S., 2013. A market mediation strategy: How social movements seek to change firms' practices by promoting new principles of product valuation. *Organization Studies*, 34 No. 5–6, 683–703. https://doi.org/10.1177/0170840613479227.

Ebit/Nielsen, 2020. Webshoppers 42 ed. https://drive.google.com/file/d/1GARLXu 6gyuDJJ8vfBU2D73ubGZIYde8N/view (Accessed 28.09.20).

Echegaray, F., 2016. Corporate mobilization of political consumerism in developing societies. *Journal of Cleaner Production*, 134, 124–136. https://doi.org/10.1016/j.jclepro.2015.07.006.

ECLAC, 2020. Observatorio COVID-19 en América Latina y el Caribe. www.cepal.org/en/topics/Covid-19 (Accessed 27.08.20).

The Economist, 2020. Covid-19 has led to a pandemic of plastic pollution. www.economist.com/international/2020/06/22/Covid-19-has-led-to-a-pandemic-of-plastic-pollution (Accessed 09.03.20).

El Issa, E., 2020. Survey: How the pandemic alters Americans' financial habits. *Nerdwallet*. www.nerdwallet.com/article/finance/covid-19-study (Accessed 19.09.20).

Fernández, R., Crivelli, L., Guimet, N., Allegri, R., and Pedreira, M., 2020. Psychological distress associated with COVID-19 quarantine: Latent profile analysis, outcome prediction and mediation analysis. *Journal of Affective Disorders*, 277, 75–84. https://doi.org/10.1016/j.jad.2020.07.133.

Fisher, M., and Bubola, E., 2020. As coronavirus deepens inequality, inequality worsens its spread. *The New York Times*. https://www.nytimes.com/2020/03/15/world/europe/coronavirus-inequality.html#:~:text=In%20societies%20where%20the%20virus,likelier%20to%20die%20from%20it (Accessed 09.03.20).

Fitzgerald, M., 2020. Many Americans used part of their coronavirus stimulus check to trade stocks. *CNBC*. www.cnbc.com/2020/05/21/many-americans-used-part-of-their-coronavirus-stimulus-check-to-trade-stocks.html (Accessed 09.03.20).

FMCG Gurus, 2020. Is coronavirus changing how we eat? www.foodnavigator.com/Article/2020/05/11/Is-coronavirus-changing-how-we-eat (Accessed 10.01.20).

Fonseca, A., Amorim, L., Cardoso, D., Ribeiro, J., Ferreira, R., Kirchhoff, F., Monteiro, A., Santos, B., Ferreira, B., Souza Jr., C., and Verissimo, A., 2020. Boletim de Desmatamento da Amazonia Legal. Belem: IMAZOM (October). https://imazom.org.br/publicacoes/boletim-do-desmatamento-da-amazonia-legal-outubro-2020-sad/ (Accessed 11.11.20).

Garcia-Priego, B. A., Triana-Romero, A., Pinto-Galvez, S. M., Duran-Ramos, C., Salas-Nolasco, O., Reyes, M. M., and Troche, J. M. R., 2020. Anxiety, depression, attitudes, and internet addiction during the initial phase of the 2019 coronavirus disease (COVID-19) epidemic: A cross-sectional study in Mexico. *medRxiv*. https://doi.org/10.1101/2020.05.10.20095844.

Geissdoerfer, M., Savaget, P., Bocken, N. M., and Hultink, E. J., 2017. The circular economy – A new sustainability paradigm? *Journal of Cleaner Production*, 143, 757–768. https://doi.org/10.1016/j.jclepro.2016.12.048.

Giddens, A., 1991. *Modernity and Self-Identity*. Cambridge: Polity Press.

GlobeScan, 2020. Global south rising: A Globescan insight. https://globescan.com/global-south-rising-insight/ (Accessed 09.02.20).

Grunert, K. G., Hieke, S., and Wills, J., 2014. Sustainability labels on food products: Consumer motivation, understanding and use. *Food Policy*, 44, 177–189. https://doi.org/10.1016/j.foodpol.2013.12.001.

Hamari, J., Sjöklint, M., and Ukkonen, A., 2016. The sharing economy: Why people participate in collaborative consumption. *Journal of the Association for Information Science and Technology*, 67 No. 9, 2047–2059. https://doi.org/10.1002/asi.23552.

Hanss, D., and Böhm, G., 2012. Sustainability seen from the perspective of consumers. *International Journal of Consumer Studies*, 36 No. 6, 678–687. https://doi.org/10.1111/j.1470-6431.2011.01045.x.

Horne, R. E., 2009. Limits to labels: The role of eco-labels in the assessment of product sustainability and routes to sustainable consumption. *International Journal of Consumer Studies*, 33 No. 2, 175–182. https://doi.org/10.1111/j.1470-6431.2009.00752.x.

Huber, B., 2016. Slow food nation: How Brazil challenged the junk food industry and became a global leader in the battle against obesity. *Food & Environment Reporting Network*. https://thefern.org/2016/07/brazil/ (Accessed 09.09.20).

IBGE, 2019. Pesquisa Nacional por Amostra de Domicílios (PNAD) Contínua: Características dos Domicílios e dos Moradores (Accessed 09.06.20).

IBGE, 2020. Pesquisa Nacional por Amostra de Domicílios (PNAD) Contínua: Características dos Domicílios e dos Moradores. https://biblioteca.ibge.gov.br/visualizacao/livros/liv101707_informativo.pdf (Accessed 09.09.20).

IFIC/FoodInsight, 2020. Survey: 85% of Americans have changed what they eat during pandemic. www.foodmanufacturing.com/consumer-trends/news/21136861/survey-85-of-americans-have-changed-what-they-eat-during-pandemic (Accessed 08.09.20).

ING, 2020. International survey coronavirus: Saving differently, not equally. https://think.ing.com/uploads/reports/ING_International_Survey_Coronavirus_-_saving_differently_not_equally_REPORT_FINAL.pdf (Accessed 09.09.20).

Inglehart, R., 1990. *Culture Shift in Advanced Industrial Society*. Princeton, NJ: Princeton University Press.

Instituto SocioAmbiental, 2020. Desmatamento e Covid-19 explodem em Terras Indígenas mais invadidas da Amazônia. https://www.socioambiental.org/pt-br/noticias-socioambientais/desmatamento-e-covid-19-explodem-em-terras-indigenas-mais-invadidas-da-amazonia (Accessed 11.11.20).

IPEA, 2020. Potencial de teletrabalho na pandemia: um retrato no Brasil e no mundo. *Carta de Conjuntura*, No. 47, 2° trimestre.

Ipsos Global Trends, 2020. Understanding complexity. www.ipsos.com/sites/default/files/ct/publication/documents/2020-02/ipsos-global-trends-2020-understanding-complexity_1.pdf (Accessed 09.09.20).

Ipsos-Mori, 2020. Public opinion on the COVID-19 coronavirus pandemic. www. ipsos.com/ipsos-mori/en-uk/public-opinion-covid-19-coronavirus-pandemic (Accessed 19.09.20).

Knickmeyer, E., Bussewitz, C., Flesher, Brown, M., and Casey, M., 2020. Thousands allowed to bypass environmental rules in pandemic. *AP News*. https://apnews.com/3bf753f9036e7d88f4746b1a36c1ddc4 (Accessed 09.07.20).

Kronthal-Sacco, R., and Whelan, T., 2019. Sustainability share Index™: Research on IRI purchasing data (2013–2018). *Center for Sustainable Business, NYU/Stern*. www.stern.nyu.edu/sites/default/files/assets/documents/NYU%20Stern%20 CSB%20Sustainable%20Share%20Index%E2%84%A2%202019.pdf (Accessed 09.11.20).

Madgavkar, A., White, O., Krishnan, M., Mahajan, D., and Azcue, X., 2020. COVID-19 and gender equality: Countering the regressive effects. *McKinsey Global Institute*. www.mckinsey.com/featured-insights/future-of-work/Covid-19-and-gender-equality-countering-the-regressive-effects (Accessed 09.03.20).

Mao, C., Koide, R., and Akenji, L., 2020. Applying foresight to policy design for a long-term transition to sustainable lifestyles. *Sustainability*, 12 No. 15, 6200. https://doi.org/10.3390/su12156200.

McKinsey, 2020. Survey: Brazilian consumer sentiment during the coronavirus crisis. May. www.mckinsey.com/business-functions/marketing-and-sales/our-insights/survey-brazilian-consumer-sentiment-during-the-coronavirus-crisis (Accessed 13.06.20).

Mercer Consulting, 2020. Pesquisa Práticas de Trabalho Flexível e Remoto. June.

Miccli, A., 2020. Tendências de Marketing e Tecnologia 2020 em Tempos de Coronavírus. *Infobase/TEC*.

Nielsen, 2020. Cenários da Vida pós-Covid: Retomar, Reiniciar e Reinventar. www.nielsen.com/br/pt/insights/article/2020/cenarios-da-vida-pos-covid-19-retomar-reiniciar-e-reinventar/ (Accessed 09.09.20).

Norris, P. (Ed.), 1999. *Critical Citizens: Global Support for Democratic Government*. Oxford: Oxford University Press.

Pamplona, N., and Cucolo, E., 2020. Com Restrições Às Compras, Taxa De Poupança Atinge Maior Nível Em Cinco Anos. *Folha de S. Paulo*. https://www1.folha.uol.com.br/mercado/2020/09/com-restricoes-as-compras-taxa-de-poupanca-atinge-maior-nivel-em-cinco-anos.shtml (Accessed 09.06.20).

Power, K., 2020. The COVID-19 pandemic has increased the care burden of women and families. *Sustainability: Science, Practice and Policy*, 16 No. 1, 67–73. https://doi.org/10.1080/15487733.2020.1776561.

Prime, H., Wade, M., and Browne, D. T., 2020. Risk and resilience in family well-being during the COVID-19 pandemic. *American Psychologist*, 75 No. 5, 631–643 http://dx.doi.org/10.1037/amp0000660.

Reid, L., 2020. Sustainable, local, ethical: How is Covid-19 changing our shopping habits? www.brandwatch.com/blog/react-Covid-19-sustainable-shopping/ (Accessed 09.03.20).

Scherer, A., and Palazzo, G., 2011. The new political role of business in a globalized world: A review of a new perspective on CSR and its implications for the firm, governance, and democracy. *Journal of Management Studies*, 48 No. 4, 899–931. https://doi.org/10.1111/j.1467-6486.2010.00950.x.

Spencer, N., 2020. The opportunity of plant-based foods in LATAM is real. *Says DuPont.* www.nutraingredients-latam.com/Article/2019/12/06/The-opportunity-of-Plant-based-foods-in-LATAM-is-real-says-DuPont (Accessed 09.03.20).

SustainAbility Trends, 2020. Sustainable consumption. From aspiration to behavior change. *SustainAbility.* https://trends.sustainability.com/2020/sustainable-consumption/ (Accessed 09.03.20).

Szmigin, I., Maddock, S., and Carrigan, M., 2003. Conceptualising community consumption: Farmers' markets and the older consumer. *British Food Journal*, 105 No. 8, 542–550. https://doi.org/10.1108/00070700310497291.

Thompson, B. Y., 2019. The digital nomad lifestyle: (Remote) work/leisure balance, privilege, and constructed community. *International Journal of the Sociology of Leisure*, 2 No. 1–2, 27–42. https://doi.org/10.1007/s41978-018-00030-y.

Torales, J., O'Higgins, M., Castaldelli-Maia, J., and Ventriglio, A., 2020. The outbreak of COVID-19 coronavirus and its impact on global mental health. *International Journal of Social Psychiatry*, 1–4. https://doi.org/10.1177/0020764020915212.

Torrente, F., Yoris, A. E., Low, D., Lopez, P., Bekinschtein, P., Cetkovich, M., and Manes, F., 2020. Sooner than you think: A very early affective reaction to the COVID-19 pandemic and quarantine in Argentina. *medRxiv.* https://doi.org/10.1101/2020.07.31.20166272.

UN News, 2020. Address 'unprecedented' impact of coronavirus on Latin America and the Caribbean, urges Guterres. https://news.un.org/en/story/2020/07/1068051 (Accessed 09.03.20).

Van Dorn, A., Cooney, R. E., and Sabin, M. L., 2020. COVID-19 exacerbating inequalities in the US. *Lancet*, 395 No. 10232, 1243. https://dx.doi.org/10.1016%2FS0140-6736(20)30893-X.

Waytz, A., and Gray, K., 2018. Does online technology make us more or less sociable? A preliminary review and call for research. *Perspectives on Psychological Science*, 13 No. 4, 473–491. https://doi.org/10.1177/1745691617746509.

Zypmedia, 2020. Consumers want to support their local economy by supporting local businesses. www.prnewswire.com/news-releases/consumers-want-to-support-their-local-economy-by-supporting-local-businesses-according-to-a-survey-by-zypmedia-301066610.html (Accessed 09.03.20).

3   New normal lifestyles

Future scenarios

Many scholars, journalists, authorities, and citizens believe that the Covid-19 crisis will leave an indelible and long-lasting mark on societies. Some expect this mark to be nothing less than an emotional rescue and embracement of old business as usual, driven by personal priorities to restore the old ways, further incentivized by contextual stimuli from business and government in need of urgent reactivation. This is the hypothesis of the back to old normal. But, for many others, a new normal is what is to be expected, riding on the emerging trends reviewed in Chapter 2 and favored by both the pro-change proponents among the general population and the various choice architectures designed by policy makers, responsible corporations, or activist NGOs to cope with the virus outbreak. What will the post-pandemic future look like? How will it affect future lifestyles?

One popular prognosis reduces future patterns to a polarity between bouncing back to old ways (usually labeled "back to normal") and the embracement of a "new normal". Simplistic and reductionist as it may sound, this polarity has caught attention worldwide, with over 90 million entries labeled as "new normal" in Google Search, as of mid-November 2020. Zooming in on the post-Covid-19 world through the filter of sustainability-related topics reveals an impressive 7 million entries.[1] Public opinion studies indicate that as the pandemic death toll reaches new highs, people may stick to a new normal not so much because of a conscious choice for embracing a sustainable future, but rather as a result of a heightened risk avoidance thinking, which may offer some indirectly attained lower footprint dividends.[2]

In Chapter 2 we identified trends – increasing, decreasing, and newly emerging – in ten key domains of practice. Some of the observed changes are temporary adjustments to the new situation and are likely to disappear when the pandemic is over. Tele-education and online entertainment and leisure at the current mass level are two developments likely to recede and to turn into behaviors of a minority. Others are more likely to last because their impact includes in many individuals changes of attitudes, beliefs, and values. Mindful consumption, localism, digitalization affecting work performance, the normalization of social isolation, and a new role for home entail more enduring, resilient tendencies highly likely to persist.

This chapter considers changes in two domains of practice identified in Chapter 2: consumption and social relations. Taking scarcity as a powerful modulator of social values, two ideal types of responses are identified for consuming and socializing, generating four different lifestyles scenarios. The study considers behavioral patterns of each lifestyle across ten areas of practice, revealing opportunities and challenges faced by organizations engaged in a pro-sustainability agenda. This analysis considers developments mainly occurring in Brazilian society, although occurrences have parallels in other Global South contexts.

Impacts on consumption and social life: changes in intensity, characteristics, and daily manifestations

The pandemic has seriously affected practices of consumption and socializing among individuals. Lockdown, social isolation, and physical distancing blocked both activities. While families whose incomes were not affected by the pandemic have actually more money because of restricted spending on leisure travel, restaurants, and other elements of discretionary incomes, the majority of people saw their personal finances shrink. The reduction of personal financial resources affected net volumes of consumption together with a substantive redirection from durable and non-essential goods during pre-Covid-19 times to essential, food staples, and home cleaning products (McKinsey, 2020). Surveys in mid-April indicated that two-thirds of Brazilians avoided expenditures in non-essential items, while an additional one-fourth intended to do likewise (OpinionBox, 2020). Mass financial hardship affected over half of total households and two-thirds of families from the lower classes (IBRE/FGV, 2020).

Similarly, stay-at-home practices significantly reduced social interaction and interpersonal connections, reducing the intensity of social life. Google community mobility reports indicate a drastic reduction of movements during the first 60 days of quarantine, with lower movements indicating fewer human interactions. Trips to recreation sites decreased by 53% 2 months after quarantine was issued, and visits to parks and outdoor spaces fell by 50%. Other typical socializing venues like workplaces recorded a fall of 26% (Ritchie, 2020). Reports by the InLoco mobility app pointed in the same direction. Light or zero social distancing was followed by one in three to four citizens (Datafolha, 2020; OpinionBox, 2020), thereby suggesting that the vast majority of the population significantly decreased their level of social contacts with others not residing on the same premises. This is even more drastic for people living in single-resident units, which in Brazil total 15% of population or about 12 million individuals, over half of them aged 60 years or older (IBGE, 2020). Increase in loneliness and isolation among Brazilians (Ipsos-MORI, 2020) reflects the spread of social distancing, which conveys an indication of lower volume of interpersonal interactions (Brooks et al., 2020).

Ways of consuming and socializing also underwent major shifts. Modes of shopping and consumption migrated from in-store, in-person to the usage of online channels and home delivery. Eating out in restaurants and frequenting cafes (which accounted for roughly one-sixth of meals and one-third of total spending in food) came to a sudden halt. Online purchasing expanded over 40% for supermarkets and drugstores (OpinionBox, 2020), with one-third of Brazilians relying on e-commerce for the first time (Comscore, 2020). Niche product consumers of organic food and direct-from-producers purchase networks migrated from street fairs and specialized shops to messaging apps (Lopes et al., 2020).

Different patterns of social relations resulted from the quarantine. Internet-mediated connections became the rule for maintaining an active social life even for those living with others, as usage of communication and video-messaging apps scaled up (Comscore, 2020). Social connectivity thus moved online with record downloads of Zoom and TikTok (Correio Braziliense, 27/3/2020). Conversely, offline socializing with the outer world remained feasible for those with balconies that occasionally played the role of an external living room to chat with neighbors. Offline social interactions occurred between those living in overcrowded slums and tenements that hardly enable social isolation.

Lastly, consumption and social companionship experienced substantive changes in terms of content, or goal fulfillment. Consumption is usually connected to utilitarian functions of needs satisfying (more clearly represented in food provisioning) and purposes of social distinction through status symbolization or signaling (Jackson, 2005). Likewise, social interaction embodies an instrument for fulfilling needs of social identity and social esteem. Under the pandemic, as consumption and social companionship become in short supply, both practices called for a revision of goals.

The role of consumption was reset by public health priorities driven by fear of contagion. Needs satisfaction was concentrated on essential foods and home cleaning products and social contacting was changed to being undesirable. Social connectivity that usually contributes to emotional and cognitive realization and wellness was transformed into a threat to personal wellbeing, as others become potential vectors of disease (Abel and McQueen, 2020). Furthermore, financial difficulties and social isolation transformed the display of status symbols to being superfluous and inconsequential. In fact, status display through conspicuous consumption became irrelevant when physical meetings did not take place. Status signaling did not stop but was adapted to the online atmosphere. Prior crises, for example, revealed shifts from the accumulation of material possessions to conspicuous modes of conservation, using eco-friendly and ethical status markers (Griskevicius et al., 2010). Conversely, stay-at-home imperatives brought new value to social life, which furthered identity reinforcement beyond social comparisons, group belonging and conformity to social norms. Thus, core goals of socializing were more aptly redefined in terms of emotional support, exchange of information, and recognition of affinities and shared values.

A scarcity theory of post-Covid-19 behavioral repertoires

Scarcity as a stressful life event represents a driver of unexpected, exogenously imposed adjustments in habits and norms affecting specific generations (Zwanka and Buff, 2020). This effect results from human predisposition to attribute greater subjective value on those experiences and practices that exist in relatively short supply (Maslow, 1943). Scarcity shapes needs and establishes priorities or values, particularly after a sudden change from a previously affluence equilibrium. Scarcity thus upholds an overwhelming socialization force capable of reshaping beliefs and behaviors (Inglehart, 1990). Literature points to numerous behavioral changes resulting from exposure to cataclysms characterized by sudden and generalized levels of scarcity. Coping mechanisms that followed traumatic events or system failures include intensified risk-averse reactions, migration from hedonism to utilitarianism as main criteria for choice, and reduced overconsumption and overspending, occasionally followed by a new practice of savings (Zwanka and Buff, 2020). Market failures in a globalized supply chain of products propelled these behavioral adaptations generating scarcity by repeated episodes of hoarding and extreme buying (Kirk and Rifkin, 2020), or by shortage of food and few prospects for supply, which stimulates the self-production of food (Colby, 2020).

Responses to scarcity in consumption and social life

The redefinition of conventional goals, procedures, and volumes of consumption and social relations makes these two the scarcest resources in society. Consequently, thinking ahead of the post-Covid-19 world, two opposing ideal-type responses are identified in each domain. Real-life situations may fall within degrees of each extreme, but these ideal types can help to map out the range of future scenarios.

In terms of consumption, one possibility is a radical back-to-old normal type of response. This embodies a "consumer revenge" type of response, where delayed gratification and repressed consumption push compensation practices consisting of the accumulation of material goods. This might fuel conspicuous consumption back into existence. The opposite response type consists of fulfilling needs through non-material goods and the valorization of frugality and sufficiency experienced during the quarantine. Previous crises have also led to a reappraisal of ways of living in favor of less materialistic priorities and a generalized reduction of consumer volumes (Kennett-Hensel et al., 2012). This choice implies the abandonment of material status markers and a rejection of the work/spend cycle treadmill, thus giving way to a "post-material frugality" type of response.

In terms of social companionships, one possibility involves a desire for physical face-to-face sociability with others. This embodies a "social diving" type of response by rushing into social and group situations, high physical out-of-home exposure aimed at recovering the human contact with others, catching up with the deprivation experience of being left without affections and positive social emotions, urgently compensating for frozen or repressed senses of identity and belonging.

The opposite response is that individuals may find satisfaction in remote social relationships, conveniently acclimatized to online interfaces for performing social life, maintaining privacy, and providing stronger benefits than the costs of solitude. Previous research has indicated that humans who feel lonely tend to use social media more and, in some cases, even prefer social media to physical interaction (Nowland et al., 2018). For this subgroup, the internet is not just a mere "add-on" with, in most cases, limited impact on the physical world, but actually a substitution or preferred alternative to the physical, real world – as worrisomely acknowledged by some scholars (Donthu and Gustafsson, 2020). This is the "internalization of virtuality" type of response.

Segmenting post-Covid-19 lifestyles scenarios

The intersection of ideal-type responses yields four hypothetical settings or scenarios. Scenarios are stories about what the future may be like (Corwin et al., 2020). These are speculative narratives about how different individuals may engage in distinctive lifestyles given the dramatic conditioning experience that they underwent in terms of consumption and social relations.

"Consumer revenge" conflated with "social diving" generates the extreme "back to normal" scenario. This lifestyle profile allocates high values to material consumerism and intensified interpersonal interactions. When "consumer revenge" combines with a response type of "internalization of virtuality", this breeds the "wireless materialists" lifestyle scenario. This profile favors a commercialized realization through acquisitions and social cocooning behaviors that minimize human connections. A third type of lifestyle scenario hosts those matching a "post-material frugality" response with a "social diving" reaction, which yields a "gregarious simplifiers" scenario. This profile embraces a low-carbon lifestyle and channels social identity through intense social bonding and community connection. Lastly, a combination of "post-material frugality" with the "internalization of virtuality" creates the "click rebels" scenario. This lifestyle deemphasizes materialism as catalyzer of self-realization while successfully transferring social life (as well as other key life practices) to virtual platforms. Figure 3.1 illustrates these scenarios.

How would these emerging lifestyles negotiate roles and goals across the major dimensions of life? What sort of behaviors can one expect regarding work and leisure, study and wellness, love relationships and social bonding, citizenship and goods provisioning, and what would be their implications for sustainability?

Table 3.1 offers a summary of the lifestyle repertoires across domains revealing the favorable and unfavorable socio-environmental effects of the emerging behavioral repertoires. Given the comparatively higher contribution of social life to subjective wellbeing, compared to returns from material consumption (Etzioni, 2011), we can expect a lower environmental footprint and more socially responsible or inclusive behaviors from those embracing more

SOCIAL RELATIONS

SOCIAL DIVING   VIRTUALITY INTERNALIZED

	SOCIAL DIVING	VIRTUALITY INTERNALIZED
CONSUMER REVENGE	BACK-TO-NORMAL	WIRELESS MATERIALISTS
POST-MATERIAL FRUGALITY	GREGARIOUS SIMPLIFIERS	CLICK REBELS

CONSUMPTION

Figure 3.1 Post-Covid-19 lifestyle scenarios.

Table 3.1 Projecting lifestyles behaviors across main domains of practice

	Back-to-Normal	Wireless Materialists	Gregarious Simplifiers	Click Rebels
Work	Partial adoption of telework conditional to convenience, status signaling, and profit.	Full adoption of telework to maximize professional potential.	Resistance to telework, favoring shortened workweeks.	Telework to enable self-realization and a freelancing ethos, dynamics. Open-source collaborator.
Family/love	Materially mediated (out-of-home consumption rituals) engagement in family life. Outsourcing of family/in-home caretaking tasks.	Poor work-life balance. Social cocooning. Use of immersive tech means for family life enjoyment	Priority to family/community around green, neighboring areas and shared meals. Active engagement in offline family routines and rituals.	Sharing digitalized rituals with family, along with more egalitarian parenthood responsibilities.
Wellness	Relaxed germ-free sanitation approach. On-site visits to group-oriented fitness centers, wellness classes.	Consumer of online fitness and wellness classes. Virtual reality gym and games.	Non-monetary based mental and physical wellness routines (hikes, family life, meditation).	Germ-concerned mentality with spiritual activities investment. Avid consumer of mental health narratives.
Leisure	Live shows, dine-outs, world travelling are a must to be posted online for display.	24×7 news consumption. Avid immersive technology user for home entertainment and online gambling.	Outdoor, natural surroundings trips as top leisure priorities. Involved in DIY hobbies and nostalgia games.	Cloud-based collaborative game playing. Virtual visits to museums, destinations, gamified meetings.

(Continued)

Table 3.1 (Continued)

	Back-to-Normal	Wireless Materialists	Gregarious Simplifiers	Click Rebels
Education	Focus on networking outcomes. Resistance to home-schooling for kids.	Continuous education instrumental to personal gains.	Opposing home-schooling. Favorable to alternative education models.	Supporters of e-learning for all publics and for improving broader skills.
Mobility	Omniscient individualized automotive transportation as priority and status symbol.	Mobility minimalists. Heavy reliance on online solutions such as delivery, virtual meetings, e-commerce, e-banking.	Commuting through walking/ biking. Favorable to exploring safe carpooling/shared rides.	Dreaming about self-driving vehicles, using micro-mobility means (individualized low-impact electric scooters/ bikes).
Healthcare	On-site visits to physicians, fitness centers, wellness classes.	Heavy users of telemedicine and diet/ weight control apps.	Voluntary, other-oriented actions as therapy through care-mongering.	Adopters of remote psychotherapy and mental wellness.
Food provision and con-sumption	Mix of online/offline buying. Relying on specialized, boutique-type shops. Animal-based diet coupled with healthy supplements and organic, certified food. High waste generation due to disposals.	Full reliance on online shopping. Dependence on industrialized, frozen food. Likely excessive eating and alcohol drinking. Delivery packaging waste generator.	Priority to purchase clubs, local businesses, and direct-to-farmers markets. Values recycled, repaired, and reused goods. Transition to more plant-based organic/fresh food. Continuous home-cooking.	Convenience-driven purchase, based online at big box stores. Food leftovers optimizers Adopter of online-guided home-cooking. Reused goods favored. Highly planned purchases only.

Housing	Living in larger apartments or suburbia gated communities. Balconies as key socializing ecosystem.	Valuation of highly compartmentalized units adapted to full home-office routines. Also, likely to move to gated communities.	Waste minimizers (using segregation, composting), rejecting plastic recipients. Valuation of green, well-preserved landscapes and naturally aired/lightened habitats. Importance to water and energy conservation.	Users of city center, small, flexible, multifunctional units. Importance to water and energy conservation.
Citizenship	Conceptually opposed to lockdown, easily engaged in Covid-19 anti-corruption protests. Political consumers through boycotts/buycotts.	Compliant citizens Likely to trade rights for tighter social control aiming at health safety. Adopters of checkbook activism.	Prosocial, self-initiated networked actions in solidarity to vulnerable groups.	Highly involved in click-activism, crowdsourcing, social media campaigning.

frugal repertoires. "Gregarious simplifiers" thus constitute the lifestyle pattern with higher sustainability credentials, followed by "click rebels". These profiles acknowledge that scarcity-triggering experiences due to the pandemic could bring enduring change toward a low-carbon, socially fair future.

Conclusions

Covid-19 affected multiple domains of practice, reshaping the spheres where individuals learn and exercise roles, responsibilities, and goal achievement to determine their needs and wants. Consumption and social relations were hit the hardest. The analysis of potential reactions once the pandemic recedes presents four scenarios, each embodying a distinctive lifestyle that finds expression in each domain of practice.

The implications of these emerging lifestyles are likely to affect the future of businesses and society's progress toward sustainability.

This chapter described consumer and social life-level changes that reshaped society's behavioral repertoires. It offered a map of what business, civil society, and policy-making organizations may find in post-Covid-19 times by analyzing how social isolation and the financial crisis recalibrated individuals' approaches to their livelihoods. In particular, it affords insights concerning impacts in 10 different and substantive domains by which individuals usually organize their livelihoods. The identification and profiling of hypothetical segments of the population enables identification of who may have high environmental footprints and who may seek social equality, as represented by opposite segments of "gregarious simplifiers" and "back-to-normal". This segmentation may help government policy makers, NGO advocacy groups, and corporate action developers and executors to focus their interventions on how to advance sustainable development goals.

Notes

1   Interestingly, a few months into the pandemic, generic searches about post-Covid depicted expectations predominantly in favor of a "bouncing back" scenario by a factor of nearly 2:1. By November, as second waves swept through Europe and Asia, searches about "new normal" exceeded those about a "back to normal" by close to 30%. (Accessed on 11/17/2020).

2   Probed about their expectations about the future, 73% of Canadians preferred a "broad transformation of society" over a mere 26% foreseeing a "back to status quo" situation *(Ekospolitics, May 7)*. Similarly, 54% of Britons hoped for changes in their own lives after the crisis and only 9% expect to be back to normal *(You-Gov/ Royal Society of Arts/The Food Foundation)*. In the US, as of mid-June, nearly 2/3 reactive to going back to business as usual (up from 57% few weeks ago) while solid majorities were still reactive to returning to workplace, external social gatherings, and out-of-home dinning *(Axios-Ipsos, June 12–15)*.

References

Abel, T., and McQueen, D., 2020. The COVID-19 pandemic calls for spatial distancing and social closeness: Not for social distancing. *International Journal of Public Health*, 65, 231. https://doi.org/10.1007/s00038-020-01366-7.

Brooks, S. K., Webster, R. K., Smith, L. E., Woodland, L., Wessely, S., Greenberg, N., and Rubin, G. J., 2020. The psychological impact of quarantine and how to reduce it: Rapid review of the evidence. *The Lancet*, 395 No. 10227, 912–920. https://doi.org/10.1016/S0140-6736(20)30460-8.

Colby, A., 2020. Solutions to the pandemic are hiding in plain sight. *New Dream*. https://newdream.org/blog/solutions-to-the-pandemic-are-hiding-in-plain-sight (Accessed 09.08.20).

Comscore, 2020. Consumo de Mídia Durante a Pandemia De Coronavirus No Brasil. www.comscore.com/por/Insights/Blog/Consumo-de-midia-durante-a-pandemia-de-coronavirus (Accessed 19.05.20).

Correio Braziliense, 2020. Zoom e TikTok São Os Aplicativos Mais Baixados No Brasil Na Quarentena. www.correiobraziliense.com.br/app/noticia/tecnologia/2020/03/27/interna_tecnologia,840692/zoom-e-tiktok-sao-os-aplicativos-mais-baixados-no-brasil-na-quarentena.shtml (Accessed 27.03.20).

Corwin, S., Zarif, R., Berdichevskiy, A., and Pankratz, D., 2020. The future of mobility after COVID 19. *Deloitte Insights* (Accessed 20.06.20).

Datafolha, 2020. Opinião sobre a pandemia coronavirus. http://datafolha.folha. uol.com.br/opiniaopublica/2020/05/1988729-60-sao-favoraveis-a-fechamento-total-para-conter-coronavirus.shtml (Accessed 27.04.20).

Donthu, N., and Gustafsson, A., 2020. Effects of COVID-19 on business and research. *Journal of Business Research*, 117, 284. https://doi.org/10.1016/j.jbusres.2020.06.008.

Etzioni, A., 2011. New normal. *Sociological Forum*, 26 No. 4, 779–785. https://doi.org/10.1111/j.1573-7861.2011.01282.x.

Griskevicius, V., Tybur, J., and Van den Bergh, B., 2010. Going green to be seen: Status, reputation, and conspicuous conservation. *Journal of Personality and Social Psychology*, 98 No. 3, 392–404. https://doi.org/10.1037/a0017346.

IBGE, 2020. Pesquisa Nacional por Amostra de Domicílios (PNAD). Contínua: Características dos Domicílios e dos Moradores. https://biblioteca.ibge.gov.br/visualizacao/livros/liv101707_informativo.pdf (Accessed 09.09.20).

IBRE/FGV, 2020. Sondagens Empresariais e do Consumidor. https://portalibre.fgv.br/estudos-e-pesquisas/indices-de-precos/sondagem-do-consumidor (Accessed 06.06.20).

Inglehart, R., 1990. *Culture Shift in Advanced Industrial Society*. Princeton, NJ: Princeton University Press.

Ipsos-Mori, 2020. Public opinion on the COVID-19 coronavirus pandemic. www.ipsos.com/ipsos-mori/en-uk/public-opinion-covid-19-coronavirus-pandemic (Accessed 19.09.20).

Jackson, T., 2005. Motivating sustainable consumption: A review of evidence on consumer behaviour and behavioural change. *Sustainable Development Research Network*, 29.

Kennett-Hensel, P. A., Sneath, J. Z., and Lacey, R., 2012. Liminality and consumption in the aftermath of a natural disaster. *Journal of Consumer Marketing*, 29 No. 1, 52–63. https://doi.org/10.1108/07363761211193046.

Kirk, C. P., and Rifkin, L. S., 2020. I'll trade you diamonds for toilet paper: Consumer reacting, coping and adapting behaviors in the COVID-19 pandemic. *Journal of Business Research*, 117, 124–131. https://doi.org/10.1016/j.jbusres.2020.05.028.

Lopes, I., Viana, M., and Alfinito, S., 2020. Redes alimentares alternativas em meio à Covid-19: reflexões sob o aspecto da resiliência. *Gestão & Sociedade*, 14 No. 39, 3750–3758. https://doi.org/10.21171/ges.v14i39.3265.

Maslow, A., 1943. A theory of human motivation. *Psychological Review*, 50 No. 4, 370–396. https://doi.org/10.1037/h0054346.

McKinsey, 2020. Survey: Brazilian consumer sentiment during the coronavirus crisis. www.mckinsey.com/business-functions/marketing-and-sales/our-insights/survey-brazilian-consumer-sentiment-during-the-coronavirus-crisis (Accessed 07.07.20).

Nowland, R., Necka, E., and Cacioppo, J., 2018. Loneliness and social internet use: Pathways to reconnection in a digital world? *Perspectives on Psychological Science*, 13 No. 1, 70–87. https://doi.org/10.1177/1745691617713052.

OpinionBox, 2020. Impacto nos hábitos de compra e consumo – COVID-19. Editions 1–14, March/June. http://materiais.opinionbox.com/pesquisa-coronavirus (Accessed 08.10.20).

Ritchie, J., 2020. Google mobility trends: How has the pandemic changed the movement of people around the world? http://ourworldindata.org/covid-mobility-trends (Accessed 09.08.20).

Zwanka, R., and Buff, C., 2020. COVID-19 Generation: A conceptual framework of the consumer behavioral shifts. *Journal of International Consumer Marketing*, 1–10. https://doi.org/10.1080/08961530.2020.1771646.

4   How Covid-19 may redefine the home

Pandemics and plagues have caused changes to homes over the centuries, especially relating to cleanliness and sanitation and the separation between public and private spaces. This chapter reviews the role of the home, as it evolved and changed over time, as a background to considering the long-term implications of Covid-19 and containment measures on the concept, functions, and role of the home. The chapter considers what constitutes a sustainable home and whether the post-Covid-19 home will be more sustainable.

The historical evolution of the home

The home has evolved from a basic role for shelter in an open multipurpose public space to a complex of separate rooms serving different functions, some public and some private. In the medieval age, the home included work, sleep, food preparation, meals, and entertainment for families, visitors, and servants, all within a common large space. The home as we know it today, as a private domestic residence for comfort and family life, first evolved in the Netherlands in the 17th century (Rybczynski, 1986). Their idea of home as a domestic retreat was not widely adopted until a much later date.

In 18th and 19th century Europe and among the rural gentry in England, the home became a place in which the wealthy displayed their possessions and entertained guests in increasingly large homes, supported by the availability of domestic staff. The major change to compact homes for the mass population evolved in the US at the beginning of the 20th century, when domestic staff were unavailable and where housewives sought labor-saving devices, convenience, and efficiency for the performance of household tasks (Rybczynski, 1986).

A major change in the concept of home took place after World War II (WWII), described by Langhamer in Britain, when the lower classes of society focused on the internal values of the home for the family. She described a home-centered society in which the husband became domesticated, the family shared leisure pursuits together, and the home provided comfort, warmth, and privacy (Langhamer, 2005). During the post-war decades, after the baby boom, families became smaller in size, there was near-universal marriage, rising affluence and consumption, a rise of domestic privacy, and a rise in TV home entertainment.

In the US, in the era following WWII, the availability of credit and mass transportation enabled a building boom described as "the American Dream". Homes represented success, displayed a reward for the scarcity of the war years, provided the owner with a clearly recognizable status symbol, and became the primary aspiration of society (Cohen et al., 2017).

Over the years, the average size of homes in the US increased by 74% or 1,000 sq. ft. between 1910 and 2010 (Muresan, 2016) and the number of people per household decreased from 4.54 to 2.58. Larger homes enabled the separation of activities into multiple rooms for different functions to which were added space for gardens and garages. The development of low-density residential suburbs or semi-rural villages was accompanied by a rise in car ownership and generated mass commuter travel morning and evening to and from the workplace.

The meaning of home

Categories of the meaning of "home" were defined by Després (1991) (as adapted by Richardson, 2018):

Home as security and control
Home as a reflection of one's ideas and values
Home as acting upon and modifying one's dwelling
Home as permanence and continuity
Home as relationship with family and friends
Home as center of activities
Home as a refuge from the outside world
Home as an indicator of personal status
Home as a material structure
Home as a place to own

Psychologists describe the home as a refuge, private space where the residents' identity is expressed (Aragones et al., 2010), to which could be added the importance of attachment and belonging (Kopec, 2012).

Chapman reviewed the role of the home through a historical perspective of its functions for providing shelter, nutrition, nurture, and social and cultural needs (Chapman 2013). He noted how changes in location, rooms, design, and furnishings displayed social status, particularly for the rich, in line with Veblen (1899), who identified it as the "conspicuous consumption" of the leisure class.

Rybczynski (1986) considered the home as an idea and concluded that the comfort of a home depended on the characteristics of convenience, efficiency, leisure, ease, pleasure, domesticity, intimacy, and privacy.

Prime reviewed the meaning of home for families. She identified the home as the social and psychological framework for the wellbeing of the family unit, providing stability and security, caregiving and support. The home provides a coherent narrative for the family unit and contributes to building resilience to cope with difficulties and unforeseen events. A well-functioning family unit builds utilitarian routines, symbolic rituals, and relational rules (Prime, 2020). The mechanics of daily life within the home provide protection, but if disrupted by an existential event, such as a pandemic, ecological disaster, or political coup, the severest effects are felt where the family unit was not functioning well within the home and where there were pre-existing conflicts, risks, or vulnerabilities. Then, stress and anxiety in the home not only are amplified but also cascade from one area of stress to another, especially where distancing cuts off external support systems (Prime, 2020).

During recent decades, the home in the eyes of millennials diverged from the concept of a unit for family wellbeing. Millennials perceive home in the context of their surroundings, neighborhood amenities, and connectivity with social groups and community. Millennials move frequently, change jobs, seek flexibility, and live in a digital world (Bialik and Fry, 2019). They seek small, functional, smart, low-maintenance homes with high-quality finish and internal amenities. They do not necessarily accumulate possessions to display "stuff", but seek experiences that they can share in person or through social media. They choose to live in locations where goods and services are easily available and where sidewalks and public spaces enable walkability, social interaction, and gatherings (Timmerman, 2015). Their concept of home has little to do with the categories defined by Després.

Millennials regard the home as providing overnight accommodation; daytime functions moved outside the home, eating out, leisure time out, and entertaining friends and family out. Millennials are willing to live in small private apartments in areas where the public and commercial spaces provide many of the functions previously performed inside the home.

The home prior to Covid-19 was already changing and had not only lost many of its functions but also lost its role as a status symbol. While older wealthy elite groups still displayed their success by purchasing larger and larger homes with increasing amenities, many millennials were not interested in being constrained by property assets and long-term debts but preferred low-asset living and rentals to enable flexibility and frequent relocation as employment became more dynamic. Home became a temporary convenient place to sleep, had no particular attachment, and could be located anywhere around the world. Some millennials became "digital nomads", moving from place to place totally disconnected to permanency of possession or attachment to community. Covid-19 found many millennials without jobs, without homes, without the support of a community, without savings, and some in debt (Hall, 2020).

The role of work in relation to the home

Historically, workshops of craftsmen were adjacent to homes, but the industrial revolution and mass production separated the workplace from the home. Some home functions moved to the workplace, such as meals and social and sport amenities, and home hours for family and leisure became those left over after work.

The dominant work pattern of long hours, little home time, and long commuter journeys did not enable people to live sustainable lifestyles. It left no time for community activities and little time for family and pleasures. Moreover, it generated huge levels of traffic congestion with consequent emissions and air pollution and required the allocation of urban space for mobility and parking. Frequent business flights, including weekly international commuting and physical attendance at conferences, added to levels of unsustainability. It was assumed that this pattern of unsustainable lifestyles would continue as long as economic growth was the dominant goal of free market economies.

Although the internet enabled remote working and virtual meetings for many years, it had not been widely adopted by employers or employees. Employers required physical presence and evaluated payment based on registered working hours. Employees sought face-to-face interaction and social connectivity. A small proportion of the population worked from home (under 4% in Israel, up to 14% in Finland and the Netherlands; ISEES, 2019). Some professions traditionally had offices attached to their homes (where permitted in zoning regulations for residential areas) such as doctors, lawyers, architects, and accountants. A growing group of millennials in knowledge-based companies worked as freelancers, able to choose their work location, whether from home or in a joint hub such as Wework.

Covid-19 suddenly turned remote working into the default for anyone who could continue to work through digital connectivity. Wherever possible, work moved back to the historical precedent of a combination with home, but the modern "home" was not designed to be a workspace.

The role of leisure in relation to the home

When homes were large, with multiple spaces inside and outside, much of leisure and pleasure time was spent in and around the home, entertaining friends and family, displaying possessions, and visiting others in their homes.

A significant part of leisure time for many was spent in private gardens attached to homes. The UK was a leader in promoting gardens, initially for the privileged landed gentry, in Victorian times seen as a way to promote moral and physical regeneration, and following WWII, as a mass urban leisure activity. "Gardens can be seen as a private retreat, a social place for sharing, a connection to personal history, a reflection of one's identity and a status symbol" (Bhatti and Church, 2010).

In the 20th century, the radio became a major focus of entertainment during leisure time in homes, later replaced by the TV, in line with the post-WWII focus on promoting family life. When families gathered around to watch TV, it provided a homogenizing experience, often used to transmit healthy lifestyles and moral values (Shaw and Dawson, 2001). It was emblematic of the modern leisure-based home at the time and promoted a consumer culture (Rees, 2019). When TV sets were available in different rooms for different members of the household, it was the forerunner of a change to "living together separately".

Home-based digital leisure transformed the role of leisure in the home. Activities could be enjoyed at low cost, without effort, with ease of access from the home. It widened individual choice of what, where, and with whom each member of the same household could choose his/her preferred experiences (López-Sintas, 2017). Home digital leisure provided access to a very diverse range of cultural and personal experiences but at the same time it strengthened the separation between individuals within the home, each with their own screen and headphones (López-Sintas, 2017).

Home digital entertainment did not replace the excitement and pleasure of leisure activities outside the home, such as sport and music events. Moreover, the demand for frequent low-cost international leisure and holiday travel increased at a very rapid rate. Leisure travel became a significant generator of emissions and exerted pressures for development on sensitive attractive natural areas. Leisure activities outside the home were increasingly seen as part of unsustainable lifestyles.

When Covid-19 forced leisure activities to be inside the home and in its immediate surroundings, it created "an opportunity for the reappraisal of leisure practices" (Lashua, 2020).

The home in Israel

Post WWII was historic for several countries who gained independence and started to build their own nation states. Israel was established in 1948 but the first years were dominated by struggles for existence. Most of the population lived in towns, many were immigrants as refugees, some were in tents or huts, and others in absorption centers, so the concept of "home" was then a question of where to find decent shelter and nutrition (Carmon, 2001). The government was strongly socialist and differences of income and equality were relatively small. The special form of "home" that developed in Israel was the kibbutz, an agricultural collective settlement, based on collective ideology, in which children were looked after in children's houses and meals were served in a communal dining room. This special model of home enabled all able adults to work and cope with difficult physical and social conditions. The kibbutz was a response to the needs and ideology of the times and has undergone almost complete privatization to separate family homes in a rural village with collective enterprises; a few retain the collective function of the dining room.

In the 1950s and 1960s, there was no "dream home"! The task of the government, with the help of Jewish communities throughout the world, was to construct as quickly as possible housing units for mass immigration, building new towns, often absorbing large families with multiple children, and providing infrastructures and public services. The result was frequently low-quality building of standard 4 level housing with few amenities. However, the incoming communities quickly re-established family and community social infrastructures, retaining, as far as possible, the traditions and culture from the countries from which they originated.

The 1970s and 1980s saw a big change in the government and the economy and consequently in the concept of "home". The government changed from a socialist to a neoliberal market economy and privatized national assets and companies, including public companies that had been operating infrastructures and developing industries based on natural resources. Inequalities began to emerge, gaps in education and income levels widened, and the population looked at what was happening elsewhere in the world, in addition to its own domestic context. Tel Aviv became a globalized business and financial center, hi-tech industries and start-ups characterized the innovative economic scene, and the successful and prosperous copied the patterns of the "home" as a reward and as a status symbol, moving to larger properties in suburbs, with conspicuous consumption and generating the commuter traffic congestion as already well known throughout many countries.

Millennials in Israel, as around the world, did not necessarily adopt the values and status symbols of their predecessors, left family homes, and moved to central Tel Aviv, which prided itself on being a city where life never stops. Young people left towns in the periphery to seek higher incomes and enjoy the vibrant street life, which has characterized Tel Aviv over the last decade.

Although the desire for suburban large homes and the move of millennials to small apartments is typical in many countries, the Israeli population has retained throughout the years very strong family and community relationships. Multi-generational families join together at festivals and continue to enjoy traditional meals, which are the central feature of what they would describe as "home". Security and safety are terms associated with the responsibility of the state. Permanence and continuity are associated with social relationships, not with the built structure in which they are currently living. This may reflect the historical cultural context that Jews have frequently had to change the location of where they could live but took their religion, culture, traditions, and family relationships with them.

The outstanding impact of Covid-19 on the Israeli home was the prevention of family gatherings. This was inconceivable at first and ran against all family and cultural instincts and traditions. The government order that prevented families joining together for the religious Passover Seder (and later for the New Year and Day of Atonement) was enforced but intensely disliked. The separation of grandparents from grandchildren was completely against all notions of family life, and the prevention of community and family gathering for prayers, weddings, and celebrations ran up against public opposition from the Jewish and Arab populations in Israel. The post-Covid-19 Home in Israel will be the reestablishment of normal family and community relationships and activity.

Box: The urban kibbutz

The kibbutz originated as a collective agricultural settlement, established by immigrants to Israel with a strong socialist ideology and as a way of coping with very challenging physical and social conditions. In the 1990s, many of the kibbutzim underwent privatization but the concept of sharing work, space, and facilities was adapted in varying ways to an urban context. There are 215 cooperative urban communities with some 20,000 members, in 100 large and smaller towns, which vary in the degree of collective or co-living (Katz, 2020). Many were established by groups who had grown up with a socialist ideology and wished to continue a joint form of lifestyle. For example, a group of 90 people aged 26–38 years established an urban kibbutz in Akko, living for the first 10 years in an abandoned building but managed to move to permanent accommodation with the help of philanthropic funds and a loan repayable over 15 years. They work individually or in groups, many in educational roles, keep their own incomes but contribute to shared and joint expenses and maintain a highly active community life including weekly Friday night dinners.

What are sustainable homes?

Although there is no single accepted definition of what constitutes a sustainable home, many initiatives have focused on the reduction of the ecological footprint of housing, both in construction and during use. Others consider lifestyles within the home, with neighbors and within a community.

The most frequently used term is "green building" for which several international and national standards have been adopted. The Leadership in Energy and Environmental Design (LEED) standard US Green Building Council (USGBC) requires resource efficiency, with a focus on energy efficiency, water saving, and recycling. Passive house certification requires a higher level of energy efficiency during the use of the building. Additional green building requirements take into account the level of embodied energy in the extraction and processing of building materials.

The green building or passive house certification relates to a single building, irrespective of its location. An additional concept relates to the building within the context of a neighborhood, which includes wider issues of sustainability, such as access to public transport and efficient use of built space (LEED neighborhood development).

Minimalists propose that a sustainable home is a small home, primarily to reduce the accumulation of "stuff". Smaller houses have a lower requirement for energy for heating/cooling, but tiny houses may not be sustainable if they are an inefficient use of urban space or built on open green space (itinyhouses.com).

The sustainability of a home may depend on its occupancy level. Single-occupancy households may have relatively small physical built space, but multiple single occupancy may result in multiple appliances, generating a relatively high ecological footprint. In contrast, large families living together in a single household may have a smaller footprint per capita than multiple one-person households.

A social sustainability approach to sustainable homes would give greater priority to community resilience, proposing that people will find wellbeing and happiness where they are part of community interrelationships, whether family, friends, or neighbors. Socially sustainable homes would include sharing spaces and amenities, such as in cohousing and co-living and to some extent in condominiums. The Urban kibbutz in Israel is a special example of social sustainability (see Box).

In the decades prior to Covid-19, concepts of common, shared, joint or public spaces, amenities, and appliances gained popularity as millennials recognized the benefits that could be gained from sharing. The sharing economy offered a way to live well at lower cost without the burden of ownership or debts, although its consequences have not always been sustainable (Schor, 2020). Collaborative consumption appealed to digital natives, who were happy to obtain goods and services when needed through online apps without the need for storage of underutilized equipment. Uber removed the need for a parking space. Sharing goods between strangers and using secondhand goods became acceptable social norms. Covid-19 may have called some of the sustainability criteria of homes into question as fears arose from close proximity and contagion.

How Covid-19 changed the role of the home

The pandemic caused a dramatic shift in routines of family life on a magnitude likely not seen since WWII (Prime, 2020). When Covid-19 arrived, it found many people completely unprepared for long confinement inside their place of residence. It disrupted the mechanics of daily life and aggravated situations of stress, loneliness, and anxiety (Prime, 2020).

"The message of 'stay home' had an implicit understanding that this is the one place we can retreat to for some semblance of safety, a place where we can control who comes and goes and so fully practice social distancing" (Byrne, 2020). Sadly, many people around the world did not even have a home to which they could retreat safely, but even for those who had, the home did not necessarily provide the essential needs of the Covid-19 era.

Small homes in vibrant city centers were not able to supply the multiple spatial needs of households in lockdown. Workspaces in bedrooms, learning spaces in kitchens, and entertainment spaces all over the place conflicted with each other, conflicted between different members of the household, and demonstrated clearly that the home had to be revalued.

Covid-19 showed that homes were important. They required multiple flexible spaces for a changing range of activities during the day and overnight comfort. It showed the need for larger kitchen space, storage space, and a connection with the outside. It showed the importance of gardens, yards, and courtyards. For many, it revealed the importance of social connectivity within and between households, and the joy of community gardens.

Covid-19 generated a basic reevaluation of lives, the home, work, and leisure. Remote work became necessary, acceptable, and even likeable, and revealed the benefits of home-work balance, avoidance of wasted commuter hours, and availability of extra leisure time. Lockdown provided extra leisure time but it was not necessarily accompanied by a reported increase in wellbeing (Roberts, 2020).

The pandemic created a moment of change: "incredibly rapid shifts of deeply ingrained habits have far reaching implications" (Markowitz, 2020). One of those shifts may be a transformation from the home as a dormitory to the home as a multifunctional physical, social, and psychological framework for a good life.

Will the post-Covid-19 home be more sustainable and will it enable more sustainable lifestyles?

In terms of the home as a building, post-Covid-19 homes may be less sustainable. The demand for larger flexible spaces for multiple activities and the recognition of the benefits of a home with a connection to the outside are generating an increasing demand for larger homes in suburban or outer areas. As remote working increases, commuting will no longer be the overriding consideration for choosing a home location and people will be able to seek cheaper, larger properties in more attractive surroundings at a greater distance from their employer. In terms of sustainability, that could have mixed consequences. The gain in terms of emissions from commuter travel is obvious and was the "silver lining" during the pandemic. However, a change in the property market could come at the expense of building new residential development in areas of open landscapes with consequent loss of ecological habitats.

The impacts on domestic energy use may be mixed. On the one hand, larger homes use more domestic energy for heating and cooling. On the other hand, the daily distribution of energy demand changed from peaks to a more continuous curve, with a peak around midday, typical of weekend domestic use (Chen et al., 2020). That could have significance for better integration of renewable energy sources into transmission systems and on domestic rooftops.

In terms of leisure, people have rediscovered some of the older lost pleasures inside the home, which may continue to bring pleasure, such as more gardening, do it yourself, cooking, sewing, board games, and jigsaw puzzles (Roberts, 2020). If such renewed happiness at home reduces leisure shopping and the demand for frequent long distance vacation travel, it will contribute to more sustainable lifestyles (Whitmarsh, 2020).

In terms of social sustainability, home as part of a community, post-Covid-19 may be far more sustainable. It has strengthened the connection to local neighborhoods, enabled time to be made available for community activities, and increased support for local businesses. It is contributing to building resilience by recognizing the role of essential services and may contribute to health and wellbeing if people continue home exercise, home cooking, and extra hygiene. The connection to be part of a community may be the most significant gain for sustainable lifestyles.

References

Aragones, J. et al., 2010. Perception of personal identity at home. *Psicothema*, 22 No. 4, 872–879. (PDF) Perception of personal identity at home (researchgate.net).

Bhatti, M., and Church, A., 2010. I never promised you a rose garden: Gender, leisure and home making. *Leisure Studies*, 19 No. 3.

Bialik, K., and Fry, R., 2019. Millennial life: How young adulthood today compares with prior generations. Pew research center how millennials compare with prior generations | Pew research center (pewsocialtrends.org).

Byrne, M., 2020. Stay home: Reflections on the meaning of home and the Covid-19 pandemic. *Irish Journal of Sociology*. Stay home: Reflections on the meaning of home and the Covid-19 pandemic – Michael Byrne, 2020 (sagepub.com).

Carmon, N., 2001. Housing policy in Israel: Review, evaluation and lessons. *Israel Affairs*, 7 No. 4, Summer, 181–208.

Chapman, D., 2013. *Home & Social Status*. London; Routledge, No. 111.

Chen, C. et al., 2020. Coronavirus comes home? Energy use, home energy management, and the social-psychological factors of COVID-19 | Elsevier enhanced reader. *Energy Research and Social Science*, 68.

Cohen, M., Brown, H. S., and Vergragt, P. J., 2017. *The Coming of Post-Consumer Society: Theoretical Advances and Policy Implications*. Routledge.www.routledge.com/Social-Change-and-the-Coming-of-Post-consumer-Society-Theoretical-Advances/Cohen-Brown-Vergragt/p/book/9781138642058.

Després, C., 1991. The meaning of home: Literature review and directions for future research and theoretical development. *Journal of Architectural and Planning Research*, 96–115.

Hall, A., 2020. Covid 19 is domesticating millennials. *The Guardian*, 14 May.

ISEES, 2019. Remote working in Israel (in Hebrew).

Katz, M. M., 2020. Are urban kibbutzim the key to maintaining post Covid 19 family life? *Architecture and Design*, 30 July.

Kopec, D., 2012. *Environmental Psychology for Design*. Bloomsbury Academic. ISBN-10:1609011414 ISBN-13:2901609011412.

Langhamer, C., 2005. The meanings of home in postwar Britain. *Journal of Contemporary History*, 40 No. 2, 341–362.

Lashua, B., Johnson, C. W., and Parry, D. C., 2020. Leisure in the time of Coronavirus: A rapid response. Special Issue. *Leisure Sciences*, 1–6.

López-Sintas, J., Rojas-DeFrancisco, L., and García-Álvarez, E., 2017. Home-based digital leisure: Doing the same leisure activities, but digital. *Cogent Social Sciences*, 3 No. 1, 1309741.

Markowitz, E., 2020. President of American Psychologists association address on 01.07.20.

Muresan, A., 2016. Who lives largest? The growth of urban American homes in the last 100 years. *Property Shark*, 8 September.

Prime, H., Wade, M., and Browne, D. T., 2020. Risk and resilience in family well-being during the COVID-19 pandemic. *American Psychologist*, 75 No. 5, 631–643.

Rees, E., 2019. Television, gas and electricity: Consuming comfort and leisure in the British home 1946–65. *The Journal of Popular Television*, 7 No. 2, 127–143.

Richardson, J., 2018. *Place and Identity: The Performance of Home*. Abingdon: Routledge.

Roberts, K., 2020. Locked down leisure in Britain. *Leisure Studies*, 39 No. 5, 617–628.

Rybczynski, W., 1986. *Home, a Short History of an Idea*. London: Penguin Books.

Schor, J., 2020. *The Problem of Work: The Promise and Perils of the Sharing Economy*. Porchlight. https://www.porchlightbooks.com/blog/changethis/2020/the-problem-of-work

Shaw, S.M., and Dawson, D., 2001. Purposive Leisure: Examining parental discourses on family activities. *Leisure Sciences*, 23 No. 4, 217–231.

Timmerman, K., 2015. Millennials and home: Understanding the needs of the millennial generation. *Their Living Environment*. http://purl.flvc.org/fsu/fd/FSU_migr_etd-9512 (Accessed 19.12.20).

US Green Building Council. LEED Rating system, v4Homes and v4Neighbourhood Development.

Veblen, T., 1899. *The Theory of the Leisure Class*. New York: Palgrave Macmillan.

Whitmarsh, L., 2020. Britons hope to keep sustainable habits beyond Covid 19. UK Centre for climate and social transformation. CAST. Cardiff University 12.08.20.

5 Post Covid-19 lifestyles around the community in China

Communities are not only the spaces in which people live, but also important actors in social governance of modern societies. How they are governed involves the adjustment of the relationship between the state and society, between the state and its citizens, and also between community members. Communities are also important fields where common, daily practices take place. Thus, communities of all kinds have long been a field of inquiry, with different focuses and research questions in ever-changing historical contexts.

The rise and fall of communities have been observed in many countries for different reasons (Wu and Hao, 2015). In 2020, functions of communities in the battle against Covid-19 were "rediscovered" in many cities around the world. The ongoing effects of lockdown and isolation turned people's attention to community interdependence and the value of nearby, local area. Community life and neighboring ties thus gain importance through two parallel processes: top-down projects set in motion by governments that were supplemented with bottom-up initiatives and engagement of non-governmental actors (e.g., China) and a convergence of bottom-up, individual, grassroots solidarity initiatives and organized campaigning in favor of community interests (e.g., Global South and Europe).

China's approach to the pandemic involved the most restrictive measures centered on top-down activating community networks, which ended up playing an important prevention role in the control of human mobility, virus spread control, and provision of all kinds of services and supports. Government-led community mobilization became critical in two ways: (1) to support part of the population unable to enforce social isolation measures due to living conditions and provide supplies to vulnerable groups and (2) to operationalize social governance by enabling a vertical communication with authorities or implementation of authorities' policies. In parallel, the important roles of volunteers, social groups, and internet-based communities were also further recognized and will be increasingly engaged in future social governance. Other societies experienced community revival as an outcome of convergence between solidarity-driven self-initiated volunteering or local networks, neighborhood-centric actions, and formal programs of social assistance and improved quality of life issued by municipal governments, NGOs, and corporations. Initiatives in major cities across Europe and the Global South illustrate these developments.

The widespread prevalence of Covid-19 has changed social practices, and until the vaccines are widely available, the best way to effectively prevent and combat the pandemic is to influence human social behavior, including home isolation, social distancing, and stringent hygiene practices. Under the condition of physical space isolation, it is necessary to strengthen the close ties with the core circle of relationships while providing extensive and diversified information and communication of occurrences within communities in order to obtain effective epidemic prevention. Therefore, the concepts of "social capital for epidemic prevention" and "community resilience" emerged and will remain an important topic for future research (Bian et al., 2020).

This chapter unveils the secret of China's approach: community-grid-based city governance. Therein, we aim to address a few key questions: How can national policy measures be implemented at grassroot communities? What are the functions of urban communities in Covid-19 prevention and control? What kinds of community governance structures exist in cities? What experiences and lessons can we learn from Covid-19 for future community development? How do communities promote sustainable lifestyles?

The rise and fall of communities in Chinese cities

Communities, in a sense of a group of people who share common identities and living space with clear boundaries, have been the most important grassroot units in China's long history of political system. Since 221 BC when Qin Shi Huang unified China and made himself the first emperor of Qin Dynasty, many of the legacies from his time are still functioning in the country today: a centralized political system including national-prefecture-county three-level structures, a common market and currency, a standardized measurement system, and a unified national script. For example, although China has the same area as the United States, it has a single time zone. However, as the proverb says "the sky is high, the emperor is far away", social order below county level has been maintained mainly within communities from rural to urban areas, such as villages, communes, production teams in rural areas, and street neighborhoods and residential districts in cities. The social ecology and governance of these communities have changed dramatically after New China was founded in 1949, the departure point for the discussion in this chapter.

Today, under the administrative jurisdiction of Chinese cities, there are urban districts and counties governing rural areas. Urban districts are further divided into street neighborhoods, which consist of a number of residential districts. At each of these levels, governmental functional organizations exist like street administration offices and residents' committees, and Communist Party branches in parallel to these. For instance, Beijing has a population over 20 million and covers an area over 16,000 square kilometers. Under its jurisdiction, there are 16 districts/counties. By the end of 2019, Beijing had more than 3,200 registered urban communities and over 3,800 rural villages. Urban communities normally host 1,000 to 3,000 households (Beijing Statistics Bureau, 2020). Similar administrative structures are found in other Chinese cities. From the 1950s to early 1980s, most of these residential districts in Chinese cities were built around a state-owned unit and houses were distributed to employees and staff of the same unit. All property rights belonged to the state. Therefore, the maintenance and management of these residential communities were part of the responsibility of that unit, which could be any types of organizations, such as university, research institute, governmental agency, factory, and military force. Those common practices, such as community utilities, facilities, waste management, and plantation, were decided and organized by the responsible unit. Although residents' committees existed in all communities, they mainly functioned as operating arms of the government and organizations for policy publicity, with limited public participation in decision making.

With the economic reform since the early 1980s, this city governance structure was challenged during the marketization process. Production and labor markets became free. People depended less on their employers for social welfare like schools, healthcare, and housing. Suddenly, almost everything became available in the markets and human mobility increased. China ran into the first round of city construction. New housing development projects provided high-rise apartment buildings to meet the needs of a rapidly increasing urban population. These commercial residential communities accommodated residents from all walks of life, without the traditional ties and common identities. Property management services were introduced, often by housing developers, to these communities to function as housekeeper for all residents. Each of these communities has a registered name, compound walls, and multiple entrances with security guards. On the one hand, housing developers and the property management services had common interests and had the advantages of information and negotiation power; on the other hand, residents are loosely connected and not self-organized. Conflicts between the residents and property management were not rare during that period. The overall trust among residents and sense of belongings was much diluted with high pace of urban lifestyles.

In response to all kinds of "city diseases", ranging from lack of sanitation, social stability, public security, environmental pollution to vulnerability to the effects of climate change, the functions of communities were emphasized by the government since the late 1980s. The concept of urban community service was first proposed in 1986. Community service refers to the public services and other material, cultural, and living services provided directly to community members by the government, via street neighborhood committees (Dai and Yuan, 2010). In 1987, the Ministry of Civil Affairs put forward the requirement of exploring the community social service system at a national symposium on urban community service work. The community service system was further explained as a service network and operating mechanism that takes the community as the basic unit, relies on various types of community service facilities, targets all residents of the community and units stationed in the community, takes public service, voluntary service, and convenient and beneficial service as the main content, and aims to meet the needs and improve the quality of life, with the unified leadership of the Party Committee, government-led support, and multiple social participation (State Council Office, 2011).

It was not until 2000 that the Ministry of Civil Affairs issued the *Opinions on Promoting Urban Community Building in the Nation* that promoted community building nationwide. In 2006, the State Council issued the *Opinions of the State Council on Strengthening and Improving Community Services*, which put forward a number of principles and guidelines for the promotion of community services. The *Plan for Building a Community Service System* (2011–2015), a guiding document on community building during the 12th Five-Year Plan period (2011–2015), proposed important tasks such as developing multi-level and diversified community services and promoting the innovation of community service mechanisms, and also proposed the *Organization Law of the People's Republic of China on Urban Residents' Committees*.

Communities have irreplaceable functions in social governance. First, more than half of the residents' time is spent in their communities, which means that more than half of the individual citizens' practices, especially the use and disposal phases of consumption practices, take place in the community. Second, the community is the basic unit of society and the grassroots unit where state governance and social governance take place; social governance requires the participation of all stakeholders in the community, and state governance needs to be implemented in the community. Third, the geographically close ties are the essence of community formation (Tian, 2020). Fourth, there is no substitute for the community when it comes to a sense of belonging. Communities are increasingly functioning as organizers and suppliers for immediate access to services, including parking, gardening, waste management, hair dressing, bike fixing, elderly and children care, sports facilities, shared public spaces, interest groups, etc.

Belonging to a community is an important manifestation of family belonging, so there is no substitute for the sense of belonging that the community brings to the individual. Smith (1991: 182) offered the best explanation of this family-community-nation value in the Chinese cultural tradition: "heart-mind" (denotes both mind and heart), *hsin*, expands in concentric circles that begin with oneself and spread from there to include successively one's family, one's face-to-face community, one's nation, and finally all humanity. In shifting the center of one's empathic concern from oneself to one's family, one transcends selfishness. The move from family to community transcends nepotism. The move from community to nation overcomes parochialism and the move to all humanity counters chauvinistic nationalism.

Recognizing the indispensable roles of communities in city governance, the Chinese government has been improving the legal and policy context for a healthy community development. For instance, the house owners' committees, which first appeared in the 1990s in the commercial housing communities, are truly autonomous organizations of residents. According to China's Regulations on Property Management (2007–2021), the duties of owners' committee members are to convene owners' meetings, sign property service contracts on behalf of the property owners, understand owners' opinions, and supervise the property service company's compliance. Another recently emerged actor is community service stations. In 2010, the General Office of the Communist Party of China and the General Office of the State Council jointly issued the Opinions on Strengthening and Improving the Construction of Residents' Committees in Urban Communities, which for the first time proposed that community workstations could be established in communities in accordance with regulations. The main purpose of community service stations is to improve the capacity of the community to provide professional services and promote volunteer spirit. The staff of comunity service stations is recruited from the public. For instance, since 2009, Beijing has been recruiting university student social workers, who greatly improved the professionalism of social service stations.

The practices in recent years have consolidated the community-grid-based city governance structure – a term that first appeared during the fight against SARS in 2003. Since then, community-grid-based governance has been gradually perfected through practical exploration in various parts of the country, and a relatively stable model has been formed. In a nutshell, the so-called urban grid governance is

> to divide the urban administration area into several grid units according to certain standards, and to make use of modern information technology and the coordination mechanism between the grid units, based on the principle of "each has its own responsibilities, complementary advantages, management according to law, standardized operation and rapid response" principle, in accordance with the requirements of government process reengineering, the grid of economic, patrol police, sanitation, urban management personnel, such as communication, collaboration, support and other functions to be allocated in the form of a system to form a new urban governance system, in order to improve the level of urban governance and governance efficiency.
>
> (Wei, 2018: 338).

The past two decades have seen a "community revitalization" movement as part of the social governance project, along the regeneration of Chinese cities. Currently, there are various types of urban communities and governance modes in Chinese cities. As Table 5.1 shows, they fall into three major categories: government-led community, government-community hybrid community, and self-governed community. At this stage, communities are in transitions toward more socialized and multi-actors-involved governance modes, along with the regeneration of cities. Governmental actors are still the most influential actors in most community governance modes, but the role of the government is weakened or changed with the rise of the others (Yuan, 2016).

Roles of urban communities in Covid-19 control and prevention in China

Before the crisis of Covid-19, the above-described community-grid-based city governance had been operating in a normal social context; then came the test of an unprecedented crisis. China's fight against the first flare-up of Covid-19 lasted for 76 days, marked with lockdown of Wuhan on January 23 to the lockdown lifted in Wuhan on April 8, 2020. This battle has five phases and different prevention and control measures were

Table 5.1 Three major urban community governance modes (Yuan, 2016)

Governance Modes		Government-Led	Government-Community Hybrid	Community Self-Governance
Background		Traditional unit-owned	Traditional unit-owned communities in transition and appearance of neighborhood offices	Community replaces unit
Government-community relationship		Community-unit in one	Partially separated	Completely separated
Community management parties		Governmental agencies and their operating arms stationed in communities	Governmental and community organizations coexist	Community organizations
Roles	Government	Leading role	Guide and support	Indirect regulation
	Community organizations	Operating arms of the government	Dual functions of administration and community services	Independent community manager
	Residents	Limited active participation	Partial participation in community management	Active participation in multiple ways
Governing approach		Administration instead of community management	Partial community governance under the guidance and supervision of the government	Completely self-governing

introduced at different moments to respond to changing situations of Covid-19 spreading.

Phase 1 (December 27, 2019, to January 19, 2020): Initial emergency respond to the "viral pneumonia of unknown cause". The first reported case of "pneumonia with unknown causes" appeared in Chinese media in mid-December. Unfortunately, the uncertainty of scientific evidence and the information gap between center and local governments prevented the immediate implementation of any significant containment measures in the initial stage of the outbreak. Since January 3, 2020, China has been regularly informing the WHO, relevant countries, and regions about the unknown pneumonia outbreak. On January 18, 2020, the National Health Commission sent a high-level expert team to Wuhan and the team confirmed the human-to-human transmission on the next day.

Phase 2 (January 20 to February 20, 2020): Initial containment of the spread of the epidemic momentum through lockdown of entire Wuhan. President Xi Jinping gave stringent instructions, stressing that the party committees and governments at all levels must put people's safety and health as the top priority, make thorough plans, take effective measures to curb the spread of the virus, timely release information, and deepen international cooperation. On January 23, 2020, the entire city of Wuhan was locked down. On February 3, 2020, the Central Steering Group mobilized 22 national emergency medical rescue teams from all over the country to build the mobile cabin hospital in Wuhan. On February 19, 2020, the number of new cured cases in Wuhan was greater than the number of newly confirmed cases for the first time.

Phase 3 (February 21 to March 17, 2020): The number of new cases in the country has gradually dropped to single digits. From March 11 to 17, 2020, the number of new locally confirmed cases per day remained in single digits throughout the country. Overall, China's epidemic peak had passed.

Phase 4 (March 18 to April 28, 2020): Decisive results of Wuhan Defense War and Hubei Defense War achieved. No new confirmed or suspected Covid-19 cases were reported on the mainland China on March 19, 2020. In the meanwhile, the rest of the world was experiencing the outbreak of Covid-19. Within China, priority was put on prevention of importation of cases from outside China.

Phase 5 (since April 29, 2020): The national epidemic prevention and control turned to the normal. The Joint Prevention and Control Mechanism of the State Council issued "the Guidance on Normalizing the Prevention and Control of the COVID-19 Outbreak". By October 16, 2020, there were in total 259 confirmed cases from mainland China, a total of 80,766 cases cured and discharged from hospitals, a total of 4,634 deaths, and a total of 85,659 confirmed cases reported.

Summarizing, in the view of Bruce Aylward, Senior Adviser to the Director-General of the World Health Organization and head of the joint mission, China has adopted an unprecedented combination of "archaic methods and modern technology" that has yielded great results and outputs. This "old-fashioned approach" refers to the "containment strategy", which includes national efforts to encourage people to wash their hands, wear masks, and maintain social distances; mass temperature monitoring; suspension of public gatherings; and calls for people to reduce their mobility. New technologies, such as big data, artificial intelligence, the Internet of Things, and 5G, have given wings to these traditional prevention and control methods, making them more efficient and convenient for the public, for epidemiological investigation, online shopping, and education, for instance. In parallel to these control measures, numerous supportive measures have been taken by all levels of government, from the central government to cities, to support affected businesses with tax reduction, vulnerable population with social aids, workers returning to work with organized transportations, online schooling, exit-entry services, etc.

Eventually, many of the anti-pandemic measures were implemented at the community level, such as mobilizing volunteers to provide information to all residents, to sterilize buildings, and to aid the needed for shopping and healthcare. The hidden nature of the virus and the danger of community transmission make communities critical battlefields for fighting the epidemic. In particular, before the development of effective drugs and vaccines to block the virus, early detection and isolation at the community level were the only ways to break the chain of virus transmission. Moreover, even as the pandemic was emerging and receding, community-based epidemic prevention and control remain the most critical instruments of reducing secondary disasters and rebuilding society. In contrast to social isolation in general, the closure of communities in the face of an epidemic tended to be somewhat more positive and was a necessary response to an epidemic. However, as far as the consequences of community closure are concerned, the mechanism by which it worked was very complex. On the one hand, the required social distancing makes it impossible to allay and share fears in a normal way; on the other hand, the role of the community organizations has never been more prominent to guarantee necessary leadership (Tian, 2020).

It became apparent that, while the severity of the epidemic varies from place to place, there is a general consensus about the essential role of community grids in containing the epidemic (Tian, 2020). The normal community grids were upgraded in no time to so-called "super grids" in response to the crisis. These "super grids" mainly consisted of grassroots governmental agencies, Chinese Communist Party (CCP) branches, civil servants, and party members, and were complemented with support and services from residents, social groups, volunteers, property management companies, etc. Together, in each of the communities, they ensured full covering of the tasks: to detect the epidemic early at the community level, to cut off the chain of virus transmission, and to maintain order and stability in the community, to identify those in need, to guarantee provision of food and necessities, to cooperate with express delivery businesses and other social groups/volunteers, to supervise compliance, to provide updated information to community members, to conduct sterilization within buildings and on the compounds, to separate and manage hygienic wastes, to control move in and out of the residential compound, to collect feedback from residents, and to inform the upper level administrative authorities (Tian, 2020).

Regarding the community mobilization and resident participation, the roles of social workers and residents are one of the concerns for the future. On the one hand, serving the entire community with a handful of social workers is almost a mission impossible. On the other hand, reports on the stress, exhaustion, and psychological instability of community workers have been many. This can be attributed to their workload and pressure, as well as to the institutional mechanisms and working methods of social workers. Ideally, community grid workers not only are the bearers of services to the community, but also need to be excellent organizers in future.

Through the test of this crisis, questions were raised regarding the roles of different actors in community governance and resilience in future. What would be the required legislations to define the rights and obligations of each actor in emergency response? What are the strengths and weaknesses of current community governance modes? What are the needs for capacity building for community resilience? It is hoped that this pandemic serves as an alarm for people living in a risk society. The pandemic presents an opportunity to rethink the relationship between the individual, the state, and society, and to reposition the paradigms and interpretations inherent in liberal and communitarian values (Han, 2020).

Community revitalization toward future

A clear vision about what we know and what we need for the future is important for community revitalization. In the long shadow of Covid-19, it is clear that life will not be back to the same normal. How can this "super grid" function in a society that needs to be prepared for the next crisis? Covid-19 accelerated the promotion of many sustainability ideas and practices relevant to communities.

By coincidence, one of the priorities for communities right after Covid-19 was to enforce household waste classification, as it was planned before Covid-19. Since the initiation of waste classification work in the 1990s, Chinese cities have experimented with varied approaches, ranging from persuasion and encouragement, waste classification publicity and education, shaming to providing economic incentives. The results were not satisfactory (Yuan, 2016). In March 2017, the *Implementation Plan of the Household Waste Classification System* proposed that by the end of 2020, mandatory household waste classification should be implemented first in 46 cities. Following few pioneering cities like Shanghai in 2019 and Beijing in 2020, waste classification has already started to be fully implemented across the country, and has gradually changed from being an advocacy to a legal obligation for residents. Again, communities are responsible for enforcing the ordinance and promoting new lifestyles. Following the enhanced positive attitude toward communities during the Covid-19 pandemic, innovative practices and progresses have been increasingly reported, demonstrating new roles of different actors in community governance (Jin et al., 2020).

It is interesting to observe the emergence of new players and social ecology in communities. Some of the communication mechanisms and the trust enhanced during the Covid-19 time played positive roles in developing solutions to various common issues. For instance, Community Making Workshop, registered as a social enterprise and led by three young social workers, has succeeded turning the Erlizhuang community in Beijing from an obsolete compound with poor infrastructures and facilities to a demonstration project for good practices of social governance within 1 year. Started as a group of volunteers to help the elderly to use smart phones in communities, they attracted more volunteers to join to improve everything needed by residents in Erlizhuang community and played roles of initiators, coordinators, and organizers. With a resident population of more than 4,000, the community has been actively exploring innovative ways of social governance and has taken the lead in promoting domestic waste classification in Beijing by initiating a pilot project on on-site treatment of food waste. The community has introduced a food waste recycling machine, which effectively solves the bottleneck problem of collecting and transporting food waste in the community. At present, the machine can process about 1,000 kg of food waste per day and convert about 100 kg of nutrient soil. The nutrient soil can not only reward community residents for their work of waste sorting and meet the needs of residents for planting flowers and plants at home, but also be used for community gardening, creating a "green micro-circulation" within the community. Old and poor buildings were renovated to create public spaces as reading room, community hall, and gyms. The community park is now theme park for waste classification. Every week, different kinds of free activities are organized for different target groups and many are carrying sustainable lifestyle ideas, such as teaching children how to recycle wastes for growing plants, promoting walking and biking, recycling used clothes, and creating community gardens. Similar models have been scaled out in Beijing through different pilot projects and blended well with urban regeneration and sustainable transition efforts for green Winter Olympic Games Beijing 2022.

To summarize, this crisis has made many people reflect on their relationship with their communities and trust in each other. It was the trust and collaboration, with the aid of technologies, that saved us from the virus. This lesson cannot be important enough for us. The positive side of the crisis is indeed reflected in those community actions that built intergenerational and interclass bridges based on spiritual accomplishment as much as social justice and progressive thinking. Around the world, top-down and bottom-up community initiatives co-existed across different societies and geographies. Often, grassroots, individual-initiated volunteer actions in solidarity with specific neighboring people or vulnerable groups made headlines and inspired other individuals to replicate these efforts. These actions address specific effects and targets of the quarantine measures: some are oriented toward the socio-psychological issues raised by social isolation, others to cope with the financial hardships of the forced stoppage in the economy, and still others aimed to deal with the new disinfestation, cleanliness principles for social order. Accordingly, those actions build intergenerational and interclass bridges based on spiritual satisfaction as much as social justice and progressive thinking.

Self-initiated pro-social and cooperative behavior took the form of mutual aid networks for picking up groceries or drug prescriptions, installing digital equipment for elderly people, and setting up telephone friendship teams around chatting and reading to reduce the feelings and reality of loneliness. From Buenos Aires to London and Madrid, the perils of social isolation affecting elder generations were confronted through solidarity and caremongering inspired actions by neighbors (EUROPA PRESS, 2020; BBC News, 2020). Initiatives aimed to neutralize the adverse effects of the pandemic on household or neighborhood finances also abound. In India, young people have self-organized on a massive scale to provide aid packages for "daily wagers": people without savings or stores. In South Africa, communities in Johannesburg have made survival packs for people in informal settlements: hand sanitizer, toilet paper, bottled water, and food (Monbiot, 2020).

Healthcare workers are another specific target of unprompted, self-organized prosocial actions. In the UK, at least 1 million people volunteer to help the public health system for its hotline services or to provide support to healthcare workers. In Wuhan, China, the suspension of public transportation inspires volunteer drivers to create a community fleet for transporting medical workers between their homes and hospitals. In Prague, free babysitting of physicians' children was quickly emulated across other cities. Another initiative, in Cape Town, led by a local group has GIS to map all the district's households, survey the occupants, and assemble local people with medical expertise in anticipation of potential saturation of hospitals' professionals (Monbiot, 2020).

Bottom-up grassroots actions provoked organized agents like corporations, governments, and NGOs to react by increasing participation, better identification of different needs, and getting better prepared for emergencies. Communities around the world, though organized in different ways, remain the most important shelters for people and change makers in city governance. Chinese experiences show the importance of vertical and horizontal communications and cooperation in cases of emergency. It also raised questions on how to balance protection of individual rights like privacy and use of information collection and tracing technologies for collective interest (Han, 2020).

References

BBC News, 2020. Coronavirus: "Si no tiene nada que comer se le lleva la comidita hecha": 7 muestras de solidaridad en el mundo ante la pandemia de covid-19. www.bbc.com/mundo/noticias-internacional-52079741 (Accessed 02.06.20).

Beijing Statistics Bureau, 2020. Beijing year book 2020. http://nj.tjj.beijing.gov.cn/nj/main/2020-tjnj/zk/indexch.htm (in Chinese) (Accessed 16.12.20).

Bian, Y. J., Ma, X. L., Guo, X. X., Miao, X. L., and Lu, X. L., 2020. The theoretical construction and behavioral significance of virus-combat social capital. *Journal of Xi'an Jiaotong University* (Social Sciences Edition), No. 4, 1–11. DOI: 10.15896/j.xjtuskxb.202004001 (in Chinese).

Dai, M., and Yuan, S. S., 2010. Review on urban community services. *Journal of Urban Problems*, No. 11, 25–33. DOI: CNKI:SUN:CSWT.0.2010-11-007 (in Chinese).

EUROPA PRESS, 2020. Iniciativas solidarias para ayudar a los más vulnerables frente al Covid-19. www.europapress.es/epsocial/cooperacion-desarrollo/noticia-iniciativas-solidarias-ayudar-mas-vulnerables-frente-covid-19-20200515120241.html (Accessed 16.12.20).

Han, D. Y., 2020. The post-epidemic era – reinventing social justice. *China Law Review*, No. 5, 43–56 (in Chinese).

Jin, M., Su, M. M., and Guo, J. J., 2020. Are you ready for Green community? *Guangming Ribao*, 26 September. https://baijiahao.baidu.com/s?id=1678843638720489337&wfr=spider&for=pc (in Chinese) (Accessed 16.12.20).

Monbiot, J., 2020. The horror films got it wrong. This virus has turned us into caring neighbors. *The Guardian*. www.theguardian.com/commentisfree/2020/mar/31/virus-neighbours-covid-19 (Accessed 16.12.20).

Smith, H., 1991. *The World's Religions*. San Francisco, CA: Harper, 182.

State Council Office, 2011. Community services system construction planning (2011–2015). www.gov.cn/gongbao/content/2012/content_2034730.htm (in Chinese) (Accessed 16.12.20).

Tian, Y. P., 2020. Building a system to fight epidemics in urban communities from the governance perspective. *Social Science Journal*, No. 1, 19–27 (in Chinese).

Wei, L. Q., 2018. *Chronicle of Contemporary Chinese Society (1978–2015)*. Beijing: Sino-Culture Press, v. 1, 338 (in Chinese).

Wu, X. L., and Hao, L. N., 2015. A theoretical investigation of community governance research abroad since the community renewal movement. *CASS Journal of Political Science*, No. 1, 47–58 (in Chinese).

Yuan, D. H., 2016. Effectiveness of community waste management modes. MSc thesis of Remin University of China (in Chinese).

6   How Covid-19 may reshape cities

The question we address in this chapter is whether cities will change in response to post-Covid-19 trends. Cities have survived and grown from strength to strength over the centuries, despite adverse conditions and crises. However, the changing role of the home and the increasing focus on neighborhoods may change the functions and design of central business districts (CBDs) and residential neighborhoods and the accessibility networks that link them internally and externally.

Sustainability and urbanization

In the pre-corona era, with a global urban population forecasted to reach 68% by 2050 (UNDESA, 2019), cities around the world were promoting the revitalization of urban centers, high residential densities, intensive commercial activities, and major improvements in mass transit. Urban planning policies focused on how to create high-quality living in high-density built-up areas to enable people to live and work in the city, to enjoy leisure time frequenting restaurants, coffee shops, bars, and street activities. Property owners and developers were offering smaller and smaller private living areas as property prices soared and as people spent much of their time outside the home, in public and commercial spaces. Environmentalists supported this direction realizing that vibrant urban areas, transit-oriented development, and high-density building with small living spaces were crucial to realizing goals for the reduction of greenhouse gas emissions and for the protection of ecological habitats from development. Pre-Covid-19 trends were leading to more sustainable lifestyles.

Many current trends at that time were in line with this goal, led by millennials congregating in city centers, bringing young skilled talent and willing to sacrifice internal living space to enjoy experiences and social activities and obtain many services through the sharing economy. The gig economy of freelance workers was growing, many of them meeting in shared workspaces that provided social interaction and flexible workspace. Job security was less important to many than flexibility, meaningful work, dynamic innovation, and social connectivity.

Remote home working was already possible, but employers and employees preferred face-to-face contact and supervision of working hours. Digital and cyber systems were already available but people still traveled from home to a workplace to use them. Online shopping and entertainment were already available, but people still enjoyed the physical interaction of visiting commercial centers, attending performances, and the pleasure of spontaneous social contact along the street.

Digitalization and connectivity through the internet were in use, but not at the level that changed behavior patterns or preferences. Prior to the pandemic, experts on cities maintained that cities will continue to be the centers of human activity and that digitalization will not bring into question the centralization of urban activity (Florida, 2020).

Implications of the pandemic for cities

The pandemic was a moment of change. Lifestyles are habitual, adopted in accordance with social norms and practice and notably very difficult to change. Moments of change can be found when people change jobs, relocate, or start a family. However, it is normally very difficult to initiate radical changes; only incremental changes to embedded existing habitual routines can be initiated, possibly through "nudges" (Thaler and Sunstein, 2017).

The pandemic, however, created a major disruption, not just a health problem, which is having significant impacts on the economy and on social relationships.

Some experts have suggested that the pandemic could bring an end to high-density living in cities, while others commented that historically plagues did not undermine the continued growth of cities. In the 19th century, cities were seen as exposed to industrial air pollution and badly affected by cholera outbreaks. High-density buildings were described in the US as "teeming tenements which breed disease". The fear of ill health and the search for healthy living favored garden suburbs (Howard, 1902; Barr and Tassier, 2020; Lee et al., 2020) with low density single or double units on large plots, which later flourished and expanded with suburban trains and then with mass production of the private car.

Fiorilla (2020) wondered whether Covid-19 will bring the end to an era of urbanization. Politico asked if this would be the death of the city (Hernandez-Morales et al., 2020). Michael Batty (2020) commented that "nothing could have prepared us for what has happened . . . there are no examples in the literature . . . it could sound the death knell for the compact city idea".

Others disagreed – "The pandemic can't stop the rise of cities. Cities are where ideas germinate and spread, and ideas drive business progress and human development" (Browne, 2020). Richard Florida dismisses the pessimistic forecast and claims that cities have bounced back from far worse disasters and the drivers for consolidation and productivity in urban centers will continue (Florida, 2020).

What is certain is that cities will change as a result of the pandemic (Shenkar, 2020).

Fulton (2020) foresaw that after Covid-19 cities will be different. He forecasted that street life would be different as the number of malls and retail stores diminish, and more space is devoted to restaurants, bars, gyms, and personal services. Much of the office space would not be needed as work would be remote; only space for face-to-face meetings would be required.

Changes that are now evolving in the post-Covid-19 era may have fundamental impacts on how cities will function and the ways people live. The combination of digitalization and the requirements for social distancing may be changing the context for how people make choices on where they live and how they live their daily lives. When lockdowns and social distancing were imposed, digitalization provided solutions, but it also put into question physical proximity, face-to-face contact, and the concentration of commercial, cultural, and social activities in urban centers (Pisano, 2020). Digitalization enabled dispersion and disrupted the traditional role of city centers.

The impacts of digitalization on the city

Digitalization shot forward at amazing speed in response to lockdowns. Those who were already in the digital age could carry on with their lives; others had to learn fast how to join them in order to maintain an income, obtain supplies, and find leisure pursuits and entertainment.

Digitalization was already recognized as a driving force for change and as a technological trend likely to have significant impacts on production systems and lifestyles. Its implications were associated with the rise of ICT, the Internet of Things (IoT), big data, and analytics, with the increasing use of artificial intelligence (AI) and robotization, and its impacts were likely to affect the supply of critical raw materials and unemployment (European Environment Agency, 2020). However, even the most up-to-date assessments did not forecast how digitalization would affect society and the economy during the pandemic.

Digitalization was slowly making more and more activities available online. Zoom was there before Covid-19, and so were Netflix and Amazon, but they had not penetrated to a majority of the population. During the pandemic, digitalization was a key to how society could continue to function. Surveys since April 2020 around the world show that people adopted online solutions, found that they met their needs, and responded that they will not go back to previous offline habits (Sneader and Singhal, 2020; Accenture survey, 2020; Boons et al., 2020).

Digitalization was the driving force that accelerated remote working, a change likely to have fundamental implications for the city, especially for the future of CBDs. There were signs that some work was moving online. Globally, 50% of hi-tech companies were already working from home and some were "digital nomads", working from anywhere in the world. Some countries had moved ahead permitting working from home. The Netherlands and Finland had 14% home working in 2019; Israel had only 3.4% of the total workforce working from home in 2016. The only ministries in Israel really interested in remote working were Transport and Environment for decreasing traffic congestion and its environmental externalities (Kessler et al., 2020). Previous research showed that remote working was not going mainstream (Mokhatarian, 2020) and that although the technological capabilities were available, organizational and social barriers slowed its advance.

The corona catapulted working from home to about 50% of the workforce in Israel and brought it up on the mainstream agenda even in the highly conservative public sector (as illustrated by its adoption in the Israel Civil Service Commission, which moved to a hybrid model). Suddenly, the opposition of employers and employees to working at home was shown to be irrelevant and both sides realized the huge advantages that could be gained from it, including higher productivity, lower absentee rates, and savings of office space, parking space, and supply of amenities and facilities at the workplace as well as a huge savings of wasted commuter journey time and cost. Both sides requested some level of physical connectivity, and many companies are now proposing to use a hybrid model, with some days at home and a day or two in company offices for meetings (or a week in company offices after several weeks home working), and to maintain a strong staff affiliation to the company. A variation of this model is a work location outside the home but within easy local accessibility by creating a distributed network of nodes, with each node able to accommodate several companies and accompanied by multiple amenities for employees (Bacevice et al., 2020). That model, he proposes, would provide a work setting that can express work values and aspirations. Whichever model is selected, the result for the city is a severe decline in the demand for office space in the CBD and a redistribution and decentralization of workspace.

The acceleration of digitalization has fundamental implications for shopping streets and centers. Retailers were already aware that retailing was going online. Businesses, especially the larger enterprises, were either already online or closing down, unable to compete in the market. However, most small and medium size businesses (small and medium-sized enterprises) continued to market goods to consumers through face-to-face contact and shopping streets were active and vibrant, accompanied by restaurants and cafes. Lockdown accelerated consumer purchasing to go online with delivery services, and consumers overcame barriers, such as fear of payment online. When economies opened up again, SMEs found that their customers did not return and many had to close down. Out of 40,000 SMEs in Israel, 10,000 closed down during Covid-19 (25%), compared to a normal closure of some 6% (Haaretz, 2020). The Ministry of Commerce and Industry realized that SMEs were not able to go online without assistance and started to provide them with grants and training.

Digitalization of retail has very significant implications for cities. The closure of businesses reduces the level of activity along high streets, undermines the role of city centers, and increases demand for warehousing with easy access for logistics and delivery services.

Digitalization of entertainment may also have significant implications for city centers. Online entertainment and online gaming accelerated during the pandemic and are likely to remain popular as convenient and affordable alternatives to live entertainment or to watching films in cinemas.

Digitalization of adult education (not home schooling which failed to fulfill the educational needs for children) could change the role of universities and colleges and their campuses, which are anchor institutions in many cities.

The decline of city centers

City centers are losing their traditional roles and functions. People no longer need to travel to offices or to shops or to entertainment. When multiple business enterprises and activities in city centers close down, restaurants and bars no longer have their clientele. Moreover, many city centers thrived on the influx of tourists, who may one day return but not until travel is safe and incomes are available. Abandoned office space in the city center is likely to undermine property values, severely reduce the demand for ancillary services to office workers, and reduce the functions of central city areas. Real estate agents reported in June 2020 that rental prices in London dropped 5%. San Francisco experienced a similar drop of 5% as demand shifted outwards. Tel Aviv is experiencing a similar trend (see Box). Like many cities, commercial space was the major source of municipal income and large areas planned for office use not yet built are now unlikely to be built.

City center retail is affected not only by digitalization but also by a loss of customers. On exiting lockdown, people refrained from purchasing or delayed their purchasing, wary of additional payment burdens and reluctant to take credit. Prior to the pandemic, people tended to spend spare earnings and saved little. Loss of income during the pandemic aroused an awareness of the need for savings, when 50% of the US population found themselves without the ability to cover 3 months' expenses (Parker et al., 2020). A survey in the UK found that respondents will now save far more than before the pandemic (Centre for Longitudinal Studies, 2020). The pandemic caused a reevaluation of the need for savings and a reduction in non-essential purchasing.

Entertainment and cultural institutions have traditionally been major anchors in a creative city. Cities offered a wide range of leisure time attractions, supported arts and culture, and enabled groups to gather together to express freedom of choice and attitudes. Hopefully, these functions will return but people may be less willing to spend on leisure and cultural activities if the high rates of unemployment and uncertainty continue.

City centers thrived on eating out, cafes, and bars. Decline in businesses, offices, shopping, entertainment, and tourism will severely reduce the demand for eating and drinking facilities. Gyms and workout studios were among additional city center ancillary services for before, during, or after work. These too have lost their clientele from a reduction of city center activities and some of their clients found alternatives by running and exercising in parks and by acquiring home exercise equipment.

Table 6.1 indicates whether trends may benefit (+) or reduce (−) demand for city center activities.

Certain groups choose to live in city centers. Millennials who contributed to and enjoyed vibrant city center life were a major group living in rented accommodation in city centers. Many are leaving, unable to pay the high rents after losing employment. Families who lived in city centers found city center small apartments cramped for the multiple home activities and are seeking more comfortable accommodation elsewhere. The main population left in city centers, without the tourists, are seniors, mostly wealthy, with incomes unaffected by unemployment. However, cities on the whole seek to diversify the age structure of their residents and do not want to be characterized as largely serving an ageing population. They will need to find how young people can be attracted back to the city center.

Table 6.1 How trends may affect demand for city center activities

Shopping	−	Online will reduce visits for purchasing
	+	Local neighborhood shops survive
Working	−	Remote working will reduce demand for offices
Education	−	Loss of students on city campuses
	+	Gain of students able to learn from home town
Entertainment	−	Online streaming
	−	Reduction of live entertainment
Transport	−	Reduced demand for public transport
	+	Reduced commuter travel at peak times
	+	Increase in cycling and micro mobility
Healthy living	+	Increased demand for sport and parks
	−	Move out of town
Leisure	−	Loss of urban tourism
Eating out	−	Loss of restaurants, cafes, and bars
Residential	−	Loss of millennials, drop in rentals
		Inability to pay rents and mortgages

− loss of city center activity; + gain of city center activity.

The pull of suburbs and rural villages

When people realized that the pandemic was not a passing event but a long-term transformation of the home from a dormitory into multi-functional space, people sought residential properties that would better meet their new needs. Demand rose for out-of-center larger properties in outer suburbs, smaller towns, and rural villages. The UK market research Ipsos MORI found in June 2020 that 44% Britons think that cities will become less attractive places to live in over the next few years (Marshall, 2020). Property values rose in rural locations by 126%. People were seeking outdoor space, local communities, and proximity to friends and family, within a possible commuter distance when needed. The UK RightMove property noted a rise in demand for homes in outer areas with gardens and space for a home office (Rightmove, 2020).

The OECD in "Build Back Better" (OECD, 2020a) raised the possibility that Covid-19 could exacerbate sprawling cities by an increase in demand for less dense neighborhoods and single-family houses due to perception of higher infection risk in dense housing. Suburbs and rural areas may pull out people able to work remotely, and attract those who have a fear of mass transit and those who seek private amenities (Florida, 2020). There may be a shift from the top cities to smaller towns and from city center to suburbs (CityLab Brookings, 2020). "We may see much more sprawl as people seek to get away from big cities to smaller towns, we may see a growth in car travel and a decline in public transport" (Batty, 2020). If a hybrid model of home/workplace becomes dominant among employers, people may choose larger houses on larger lots up to 80 km/50 miles out of town (Kotkin, 2020). The key factor in choosing a home location will be a high-speed internet infrastructure rather than a high-speed road.

When taken to the extreme, remote working and remote meetings could generate a significant rise in "digital nomads", working from wherever in the world they choose and moving from place to place. Bermuda, Barbados, and Estonia are offering special visas for long-term stays (Outsite blog, 2020). Theoretically, companies could choose staff from anywhere in the world that has a high-quality internet connectivity.

Urban accessibility and mobility

Covid-19 may have caused a monumental reevaluation of accessibility, mobility, and transport. Pre-Covid-19, transportation and telecommunications were considered as separate systems that did not overlap or provide alternatives to one another. The spatial distribution of land use and activities was an integral component in the assessment and prediction of future demand and in the consideration of alternative modes of transport, but the modes were physical (road, rail, bicycle, walking) and operational (public, private). Covid-19 added a totally new conceptual framework by replacing physical modes with digital modes. Accessibility became dependent on the availability and speed of the internet connection, not on the highway or metro network. The concept of exchanging physical and digital accessibility had been raised by transport experts (Lyons, 2020) but had not penetrated into mainstream transport planning. Moreover, responsibility for physical transport and for telecommunications is usually in different government ministries who do not consider them interchangeable.

Covid-19 demonstrated very forcefully that what concerned the user was the ability to access goods and services and that digital accessibility provided a far more efficient, convenient, and cheaper solution than physical accessibility.

If the digitalization of working from home, e-commerce, and digitally mediated leisure and entertainment continue (and there is every reason to think they will), the question will not be whether to travel by metro or by car but whether to travel at all.

During lockdown, use of public transport dropped by as much as 80% (Wintle, 2020; Zhang, 2020) with a consequent monumental shake up of urban mobility. The financial viability of public transport operators was undermined. Compliance with social distancing and hygiene raised costs of operation and diminished usage decreased revenues (Schmidt, 2020).

Pre-Covid-19 mass transit systems had focused on efficiently bringing masses of commuters from suburbs and outlying areas to employment and commercial activities in the city center CBD. However, if remote working and e-commerce change the distribution of activities, travel patterns, and volumes, origins and destinations will change from a star-shaped mass transport network to a single center to accessibility and connectivity between multiple distributed nodes. A model of Los Angeles showed that if remote working increased to 33% (a figure more or less in line with many predictions, including OECD), people will move to more affordable properties in the periphery, thereby reducing commuting times, and the total daily travel distance would fall by 29% (Delventhal et al., 2020). Circulation networks may now need to serve a new mobility pattern and some infrastructures may become obsolete. Vehicles, micro-mobility, and pedestrians can easily adapt by changing routes but fixed lines for train, metro, and light rail lines have permanent infrastructures, which cannot adapt to changing travel patterns.

New urban accessibility patterns may be very different from pre-Covid-19 systems. Digitization will remove a very significant level of demand from physical commuter and shopping transport and increase the demand for fast optic fiber networks. It is also likely to require a redesign of physical urban transport networks from a focus on the city center to providing accessibility between multiple nodes and clusters throughout the urban area. Schmidt (2020) suggested that the new city could be conceived and organized as a network of villages. Future transport policy may move from investing in massive infrastructures to managing mobility and providing accessibility (Audenhove et al., 2020; Hausler et al., 2020).

The question then arises: "Who is responsible for the new networks and infrastructures?" Transport physical infrastructures have largely been built by public bodies or in private/public partnerships. Digital infrastructures have to date largely been the initiative and operation of private companies. The forced move from physical to digital infrastructures as a result of Covid-19 demonstrates that this is not an item for commercial competition on the private market for profit but an item of public responsibility that should be supplied in a similar way to other public utilities and services, such as electricity, water, road, and rail transport. The private market alone will supply optical fiber fast internet networks where the demand assures a high return on investment. Public intervention is essential to ensure that every resident, whether in the center or the periphery, in urban areas or rural, and in high- or low-income groups, has affordable access to fast internet infrastructure.

Implications for urban policy and planning

In contrast to Israel, the US during the 1950s to 1970s witnessed a massive exodus "white flight" of middle-class families to suburbs, many of which were newly created during the post-war economic and population boom. The declining city centers became impoverished and ridden with crime. It took several decades, starting in the 1980s, for cities to rebound as prosperous (and expensive) economic and cultural centers.

The pandemic may have brought current urban policies into question. If post-Covid-19 trends favor relocation to low-density suburban or rural living, cities may lose the momentum of urbanity and move to higher levels of emissions from increased energy use in larger properties and in private car travel. Empty offices and retail spaces in city centers raise the issue not only of what city centers look like but also of how existing built structures can be reallocated and redesigned for other uses. The aftermath of the pandemic will require a redesign of the urban context to meet post-corona requirements and lifestyles. The following ideas may contribute to a dialogue on where we need to go.

The home – residential units

If the response of the property market generates relocation of the wealthier population to larger homes in suburbs or more distant locations, it would be disastrous for urbanity, and would generate development pressures on "greenfield" sites at the expense of open natural landscapes.

Is it possible to find solutions for a post-Covid-19 lifestyle within a high-density urban fabric?

At the level of the dwelling unit, flexible divisions of internal space would be needed, using movable and flexible partitions, pull-up and pull-down furnishings operated by control systems, the redesign of space to serve different purposes for different people at different hours of the day and night (Ben Eliezer, 2020). Each part of the internal space will need multiple acoustic partitions and communications systems to enable parallel quiet, online working, online education, online shopping, and online entertainment at different schedules as activities are at staggered times for different residents in the household.

Wherever possible, each residential unit would need a private outdoor space, such as balconies, patios, surface and rooftop yards, and gardens, which become crucial for connecting with nature and with the outside world during lockdown phases.

Ventilation systems within buildings may need to undergo revision to prevent transmission of viruses through enclosed air circulation systems. Circulation systems in high-density high-rise building will need to be redesigned with separate intakes. Fortunately, technological advances in the construction of highly energy-efficient houses will be able to satisfy that need.

A critical issue in high-rise residential building may be the entrance/exit and elevator systems. It may become possible to enable top-down access from rooftops as well as adding bottom-up access from the surface and even midway access bridges across rooftops or at high levels. Pedestrian access and circulation systems for high-rise buildings may require rethinking. Circulation patterns are also a relevant issue for delivery services. Covid-19 accelerated the role of personal delivery services to each residential unit, which may now need a delivery box, larger than a postbox, at surface level or from the rooftop (drone delivery).

Groups of residential units

Pre-Covid-19 marketing of residential homes largely focused on the individual unit, internal space, the view, and the interior design. Post-Covid-19 widened considerations to include joint spaces and the benefits of groups of residential units as opposed to separate units on individual plots. Lockdown focused attention to what was available to residents within 100 meters of their homes. Private and joint spaces became crucial, however large or small, from the lobby to the entrance path, from sitting space at the entrance level or on the rooftop to larger gardens, from a parking space or cycle rack to the adjacent roadway, and from a common entrance to multiple joint amenities. These joint areas had been eroded over the last years by contractors seeking to maximize internal space for property value and by residents seeking to reduce areas in need of management or joint management, which could result in conflicts between residents.

Post-Covid-19 may raise the value given to living in groups of residential dwellings with shared facilities and amenities in different joint and private combinations, such as condominiums, cohousing, and shared housing (see Box).

Zoning and land use designation have in the past separated residential and commercial uses of land and buildings. That is questionable today. The need for some level of contact and separation may generate demand for a combination of residential and office space within the same building complex, providing a separation of home and work but within the immediate area and with a separate entrance.

Group planned residential units could offer many advantages to their residents but they may undermine an essential characteristic of the city as a mosaic of adjacent and different communities, where people move around freely and frequently between districts along streets. The creation of grouped units with boundary separation may turn into "gated communities" of like-minded residents, where those who belong are welcomed and those who do not are kept out. Walls, fences, and restrictions on free access are completely opposite to urbanity, which seeks to increase areas for common management.

The community and the neighborhood

Lockdown turned attention to the local scene and to what is available in the immediate vicinity. It demonstrated the importance of planning neighborhoods that are self-sufficient in providing essential shops and services and provide "green spaces" for walking and sport. "Local" also brought out the best in people, prepared to help in such essential jobs as deliveries to local older residents. It could now gain further attention as neighborhoods become trusted areas for social support and as time once wasted on commuting becomes available for participating in community activities.

Neighborhood shops and restaurants are surviving the sudden changes in consumer behavior patterns as a result of the post-Covid-19 lifestyles. They may even be the main beneficiaries as livable neighborhoods attract the kinds of activity that had previously taken place in city centers (Newman, 2020). Neighborhoods could now offer distributed work nodes, providing a solution for small homes without space for remote working and generating a new demand for neighborhood businesses and services.

The social and environmental concept of neighborhood-level services for a higher quality of life became popular with the idea of a "15-minute city" (Moreno, 2020) and was adopted by the mayor of Paris, Anne Hidalgo. Mixed-use neighborhoods or small towns offering walkability and micro-mobility, and a wide range of facilities and amenities, with neighborhood remote working opportunities may replace the dormitory suburbs, car-centric and intense city centers of many cities today. Cities may then become polycentric, with a mosaic of vibrant neighborhoods. This would be a major strategic change from strict separation of residential land use and from the concentration of work places and commerce in designated areas.

Box: Venn City – a new way of neighboring

In 2016, three partners (two of them from a kibbutz) founded in South Tel Aviv a neighborhood development project designed to provide fair housing for young people, which connects not only with people in the residential area but also with businesses and residents in the Shapira neighborhood. It offers private and shared furnished apartments for short- and long-term rental together with multiple common areas spread around the neighborhood offering spaces for social and community activities. The apartments have separate or joint facilities, such as laundry rooms and kitchens, and the common areas include workspaces, meeting rooms, a music room, a tool library, and rooms for arts, meditation, and personal care (Venn city Shapira homepage). It is not a collective inclusive community with pooled income or co-housing with common meals every week. The project is not a closed exclusive "gated" community, such as Andromeda Hill in nearby Yafo (Rosen and Razin, 2009) but has a program of shared events and experiences with its own renters and the inhabitants of the neighborhood, who can join the common areas and activities for a small monthly fee, which is contributed to the neighborhood. The project has created linkages with over 35 local businesses, which can be contacted through its Venn app, to encourage the local economy (Venn City, 2020).

Venn City is basically a real estate company that buys old properties in upcoming neighborhoods, renovates them, manages them, and rents them. It appeals to young professionals and is criticized as promoting gentrification of a neighborhood (Haaretz, 2020). Its business model is housing or living as a service. This model has obvious attraction to millennials and could also be a good model for seniors, with some modification of the shared spaces.

Venn City has also established projects in Berlin and Brooklyn (New York) and hopes to establish many more around the world. The company is backed by Angel and Asset investors.

City centers

City centers will have to recreate themselves. Many existing buildings and functions will become irrelevant or scaled down, releasing built space within the existing urban area. Some office buildings may become redundant, and shopping malls may be abandoned. Eating out and entertainment may be severely cut down. Opportunities arise for the redevelopment of the urban fabric but will need considerable public investment as well as private property investment. Suggestions include transforming abandoned office buildings into affordable housing. In fact, the city center could be transformed into a neighborhood with mixed uses.

Implications for urban sustainability

Will the post-Covid-19 city be more sustainable?

There is no commonly accepted definition of what constitutes a sustainable city or indeed of what constitutes sustainability in cities. The evaluation of cities in relation to sustainability is based on a wide range of economic, social, environmental, and cultural parameters.

The following sections consider the characteristics of five frequently used concepts for sustainable cities (green, resilient, smart, sharing, and healthy) and the role these would play to alleviate the hardships imposed by a pandemic. Some characteristics are overlapping among the five concepts, while others are more specific to each idea of a city. Many would have made life in the cities easier during the pandemic.

Green city

A green city aims to reduce its footprint and provide green spaces and parks for the benefit of its residents. Green cities are usually associated with a high level of willingness of residents to participate in promoting public commitments, such as waste separation and recycling. Community connectivity is high and the achievement of environmental goals emphasizes the sense of "place". A green city is a city that pays attention to the conservation of the built environment, to historic heritage, and to the encouragement of cultural diversity.

A green city is likely to require green buildings and impose strict regulations on activities that may generate pollution, require zero or near-zero waste, recycling and composting, the use of renewable energy, and the use of environmentally friendly goods and services.

Green cities could offer attractive local parks and gardens, especially during lockdowns, when people are restricted to access within local neighborhoods.

Resilient city

A resilient city is capable of coping with shocks and crises as well as adapting to slowly changing conditions and is able to harness new opportunities. Policies include awareness of possible risks, identification of vulnerable communities, and incorporation of precautionary and preparatory measures to enable its citizens to cope with stress and distress. It ensures that its infrastructures are capable of supplying services and coping with disruption caused by a sudden event, such as flooding, and requires all built structures to be prepared for adaptation to climate change. It is an inclusive city, which supports individual and community networks for adapting to changing circumstances and for providing a safety net for its vulnerable communities in times of need. A resilient city encourages diversity. It welcomes cultural differentiation and promotes acceptance and cooperation that cut across cultural boundaries

Smart city

Smart cities are often associated with high-density urban residential developments that enable the efficient provision of public services and the use of big data networks to enable the identification, surveillance, operation, and management of public infrastructures. Smart city systems monitor and detect risks, changes, or breakdowns in service provision and provide immediate solutions. They provide open access to data and enable job mobility and freelancing. Their highly efficient infrastructures are economically cost-saving, prevent loss or waste of resources, and enable efficient use of space and buildings and the efficient operation of transportation, water, and energy supply systems.

Smart cities offer systems that enable remote working, remote education, and remote entertainment. They could enable residents to work from home or from public spaces and parks while maintaining social distancing. Smart cities also enable surveillance to support tracing and containing the spread of infection.

Sharing city

The sharing or collaborative city promotes the provision of services in preference to the purchase of goods. Sharing space, workspace, equipment, transport, and data and information (open source) are all manifestations of the basic concept that living well does not necessarily depend on the ability to purchase property and goods (Botsman, 2010) It promotes the development of trust between multiple providers and users, new ways to cope with issues of liability for damage, and new types of regulation to enable multiple users at different times of the day or week. More intensive use of existing facilities delays or even reduces the need for new facilities. The sharing city requires geographical proximity to enable easy and convenient access to services and requires a high level of digital mobile devices to enable a very high level of connectivity between residents. Sharing contributes to sustainability when it creates relationships between people and generates trust between those sharing the same item or service (as promoted by "shareable").

Sharing cities could not offer a solution during the pandemic. No one had anticipated that social distancing and the fear of contamination would put the whole idea of sharing into question. People abandoned shared transport and avoided anything with shared surfaces or unknown users.

Healthy city

Healthy cities encourage physical activity to combat the common symptoms generated by an over-sedentary lifestyle. They encourage getting to work by foot or bicycle and engagement in sport and physical activity for all ages, and pay particular attention to ensuring a high level of social contact for ageing populations. Healthy cities require fast and easy access for all to medical services. Healthy cities cannot tolerate pollution and impose severe restraints on activities likely to generate air pollution, noise, or toxic substances. They promote clean urban transport and public access to easily accessible and useable parks.

Box: Case study: Tel Aviv

"In the blink of an eye, Tel Aviv-Yafo transitioned from its worldwide reputation as 'the Non-stop city' to become 'the city that stopped' . . . The spread of the coronavirus and the subsequent directives issued by the government have had a devastating effect on businesses throughout the city" (Tel Aviv-Yafo Municipality, 2020).

Lockdown was imposed over the whole country from March 15 for 2 months, gradually exited for a month, and then reimposed as a partial lockdown from July 17 following a surge in identified cases. The medical system was well able to cope with serious cases, well within the capacity for treatment, and the number of deaths was relatively low, totaling 2,924 (December 2020) in a population of 9 million. However, the economic crisis overwhelmed the country, throwing over a million into unemployment or paid or unpaid leave, and causing thousands of small and medium businesses to collapse and close down. By the end of 2020, 20% of private businesses are likely to close (Natanzon, 2020). Tel Aviv, as the business center of the country, was the hardest hit.

Tel Aviv is a thriving vibrant city of 450,000 residents and approximately of a million, if several municipalities in the wider Tel Aviv area (commonly known as Gush Dan) are included. It attracted another 350,000 commuters every day and 2 million tourists a year. It branded itself as a city alive day and night, full of young people, bustling streets, well known for its liberal cosmopolitan atmosphere, culture and entertainment, loved for its street scene, restaurants and bars. Strategies for the city emphasized the benefits of living close to work, of using public transport, and that living in small apartments was acceptable since many services were provided in public or commercial spaces. Families with children as well as young singles chose to be part of the lively Tel Aviv urban lifestyle. It was the leading example in Israel of urbanity and sustainable lifestyles. It joined global smart, resilient, sharing, and healthy cities programs.

A total of 70,000 (mostly young) people were employed in its 2,500 restaurants, cafes, and bars, many of which did not reopen after the first lockdown. Of its 73,745 businesses, 1,000 closed down in the first wave; 16% closed by July 2020. Estimates keep rising – by July, economists talked about a closure of 20% or 85,000 businesses around the country and unemployment (including furlough) was 21%. Along the main popular shopping streets, Dizengoff and Ben Yehuda, 25% of the retail businesses closed (The Marker, 2020). Office rentals in Tel Aviv dropped by 6% and commercial rentals by 10% (CBRE, Q2 2020). Google Community Mobility reports for December 7 showed that transit use in Tel Aviv was down by 39% and workplaces were down by 29%.

Deputy Mayor Zippi Berger in an interview to The Marker expressed fears that the long-term implications of the coronavirus may have a profound impact on the city's population, some of whom may relocate and use remote working from an out-of-town location, which could leave the city with an even higher proportion of older people (The Marker, 2020).

The closure of restaurants imposed a major change on the eating habits of Tel Aviv-Yafo residents (Tel Aviv-Yafo Municipality, 2020). People did not return to former habits after exiting the first wave. A national online survey of 1,600 respondents in May 2020 showed that there would be a strong continuation of the reduction of use of restaurants

(56%) and that people will not return to shopping as before (52%) or prefer open street shopping areas to malls (57%) (National Knowledge and Research Center for Emergency Preparedness, 2020).

Tel Aviv is the leader of Forum 15, the group of wealthier cities in Israel, which recognized that cities would not return to normal, and stressed that recovery should strengthen resilience, local economy, community wellbeing, and the availability of services and facilities within neighborhoods (Forum 15, 2020).

The government ministry responsible for land use planning recognized the risk that remote working and distancing would undermine urbanity (Planning Administration, 2020), that city high streets would lose retailing, and that the fear of using public transport would increase the use of private cars.

Tel Aviv, as other cities, lost multiple sources of income (local rates were cancelled or reduced, parking fees waived) and at the same time had to cope with major extra expenses to meet new health requirements. The strategic planning unit of the municipality is reevaluating its current strategy in light of post-pandemic developments and loss of municipal budgets. If cities move toward distributed clusters, Tel Aviv may have to rethink its future form and structure.

Will the post-Covid-19 city be more sustainable?

Pre-Covid-19 emphasis in city planning and management often focused on the ability to attract employers and companies to generate job opportunities with high income levels. Municipal income was then harnessed to provide high-quality infrastructures and services to enable efficient accessibility and functioning.

The emphasis may now be on how to attract residents who could live anywhere and who will be looking for high-quality living at reasonable cost. The focus has moved to people, enjoyment of surroundings, and availability of leisure activities and services. Cities that already adopted a green, resilient, smart, sharing, and healthy agenda may be those that will now be seen as being attractive as well as being more sustainable.

City responses to Covid-19 have been diverse (OECD, 2020b; UN Habitat, 2020; UNESCO, 2020; Kaplowitz et al., 2020) Mixed-use neighborhoods or small towns offering walkability and micro-mobility and a wide range of facilities and amenities and with neighborhood remote working opportunities may replace the dormitory suburbs, car-centric and intense city centers of many cities today.

The C40 group of cities are promoting the concept of the "15-minute city" as the post-Covid-19 city organizing framework (C40 Cities, 2020a and b):

> It will help cities to revive urban life safely and sustainably in the wake of Covid-19 and offers a positive future vision that mayors can share and build with their constituents. More specifically, it will help to reduce unnecessary travel across cities, provide more public space, inject life into local high streets, strengthen a sense of community, promote health and wellbeing, boost resilience to health and climate shocks, and improve cities' sustainability and livability. The core principles of a 15-minute city are:
>
> * Residents of every neighborhood have easy access to goods and services, particularly groceries, fresh food, and healthcare.
> * Every neighborhood has a variety of housing types, of different sizes and levels of affordability, to accommodate many types of households and enable more people to live closer to where they work.
> * Residents of every neighborhood are able to breathe clean air, free of harmful air pollutants, there are green spaces for everyone to enjoy.
> * More people can work close to home or remotely, thanks to the presence of smaller-scale offices, retail and hospitality, and co-working spaces.

Services such as community-scale healthcare and education, essential retail such as groceries and pharmacies, and parks for recreation and more need to be decentralized and present in each neighborhood.

The pandemic offers cities an opportunity to adapt and transform to meet changing needs, but some trends may intensify social inequality. Favorable neighborhoods will attract investment, facilities, business centers, commerce, and entertainment. Less favorable neighborhoods may not be able to provide the ideal 15-minute city for their residents, may lose accessibility from lack of public transport services, and may lose opportunities for working in city centers. Digitalized working populations will choose the favorable neighborhoods; non-digitalized working populations may lose physical accessibility, city center services and activities, access to commercial and entertainment facilities, and educational opportunities.

Attractive neighborhoods may flourish but the city center may not. The pre-Covid-19 city, with all of its problems, may have been on a more sustainable trajectory.

References

Accenture survey, 2020. How COVID-19 will permanently change consumer behavior. COVID-19: Fast-changing consumer behavior | Accenture, April (Accessed 19.12.20).

Audenhove, F. J. V. et al., 2020. The future of mobility post-COVID. Arthur D Little, 4th edition, July. The future of mobility post-COVID | Arthur D Little (adlittle.com) (Accessed 19.12.20).

Bacevice et al., 2020. Reimagining the urban office. *Harvard Business Review*, 14 August. Reimagining the urban office (hbr.org) (Accessed 19.12.20).

Barr, J., and Tassier, T., 2020. Are crowded cities the reason for the COVID-19 pandemic? Placing too much blame on urban density is a mistake. Are crowded cities the reason for the COVID-19 pandemic? – scientific American blog network (Accessed 19.12.20).

Batty, M., 2020. The coronavirus crisis: What will the post pandemic city look like? *Environment and Planning B Urban Analytics and City Science*, 47 No. 4. The coronavirus crisis: What will the post-pandemic city look like? (sagepub.com).

Ben Eliezer, E., 2020. Living units in routine and emergency times. Studio XS report to Israel planning administration (in Hebrew).

Boons, F., Burgess, M., Ehgartner, U., Hirth, S., Hodson, M., and Holmes, H., 2020. *Covid-19, Changing Social Practices and the Transition to Sustainable Production and Consumption*. Version 1.0. Manchester: Sustainable Consumption Institute. (PDF) Covid-19, changing social practices and the transition to sustainable consumption and production (researchgate.net) (Accessed 19.12.20).

Botsman, R., and Rogers, R., 2010. *What's Mine Is Yours: The Rise of Collaborative Consumption*. New York: Harper Collins.

Browne, A., 2020. Will COVID-19 end the era of urbanization? *Bloomberg*, 22 August.

C40 Cities, 2020a. How to build back better with a 15 minute city. Top mayors pledge to build 15-minute cities for COVID-19 recovery – Streetsblog USA (Accessed 19.12.20).

C40 Cities, 2020b. Global mayors COVID-19 recovery task force – statement of principles C40: The global mayors COVID-19 recovery task force (Accessed 19.12.20).

Centre for Longitudinal Studies UCL, 2020. COVID-19 data from five national longitudinal cohort studies user guide – COVID-19 survey in five national longitudinal studies (ucl.ac.uk) (Accessed 19.12.20).

CityLab Brookings – article on website. 9 April 2020 CityLab – article on website. 9 April 2020 CityLab – article on website. 9 April 2020 CityLab – article on website. 9 April 2020 CityLab – Bloomberg (Accessed 19.12.20) Bloomberg (Accessed 19.12.20) Bloomberg (Accessed 19.12.20) Bloomberg (Accessed 19.12.20).

Delventhal, M. et al., 2020. How do cities change when we work from home? (Unpublished).

European Environment Agency, 2020. State of the European environment. Chapter 1: Assessing the global-European context and trends. The European environment – article on website. 9 April 2020 CityLab – article on website. 9 April 2020 City-Lab – article on website. 9 April 2020 CityLab – Bloomberg (Accessed 19.12.20) Bloomberg (Accessed 19.12.20) Bloomberg (Accessed 19.12.20 state and outlook 2020 – article on website. 9 April 2020 CityLab – article on website. 9 April 2020 CityLab – article on website. 9 April 2020 CityLab – Bloomberg (Accessed 19.12.20) Bloomberg (Accessed 19.12.20) Bloomberg (Accessed 19.12.20 European environment agency (europa.eu) (Accessed 19.12.20).

Fiorilla, P., 2020. Will COVID-19 end the era of urbanization? Commercial property executive. 29 July. Will COVID-19 end the era of urbanization? (cpexecutive.com) (Accessed 19.12.20).

Florida, F., Rodriguez-Pose, A., and Storper, M., 2020. Cities in a post Covid world. Papers in evolutionary economic geography # 20.41. peeg2041.pdf (uu.nl) (Accessed 19.12.20).

Forum 15: Association of Towns in Israel, 2020. Urban resilience in regular and emergency times, June (in Hebrew).

Fulton, W., 2020. How the COVID-19 pandemic will change our cities rice kinder institute for urban research 29 March 2020. How the COVID-19 pandemic will change our cities | The kinder institute for urban research (rice.edu) (Accessed 19.12.20).

Hausler, S. et al., 2020. *The Impact of Covid 19 on Future Mobility Solutions.* McKinsey. https://www.mckinsey.com/industries/automotive-and-assembly/our-insights/the-impact-of-covid-19-on-future-mobility-solutions (Accessed 09.03.21).

Hernandez-Morales, A. et al., 2020. The death of the city Politico: 27.7.20 The death of the city – article on website. 9 April 2020 CityLab – article on website. 9 April 2020 CityLab – article on website. 9 April 2020 CityLab – Bloomberg (Accessed 19.12.20) Bloomberg (Accessed 19.12.20) Bloomberg (Accessed 19.12.20 POLITICO (Accessed 19.12.20).

Howard, E., 1902. Garden cities of tomorrow garden cities of tomorrow: Ebenezer Howard: Free download, borrow, and streaming: Internet archive (Accessed 19.12.20).

Kaplowitz, G. et al., 2020. Covid 19 Impacts on suburbs and cities. University of Oregon website Urbanism next.

Kessler, N et al., 2020 The implementation of remote work policy in Israel - Experts' opinion The Israel Society of Ecology and Environmental Sciences (in Hebrew). https://www.isees.org.il/wp-content/uploads/2020/04/%D7%99%D7%99%D7%A9%D7%95%D7%9D-%D7%9E%D7%93%D7%99%D7%A0%D7%99%D7%95%D7%AA-%D7%A2%D7%91%D7%95%D7%93%D7%94-%D7%9E%D7%A8%D7%97%D7%95%D7%A7-%D7%91%D7%99%D7%A9%D7%A8%D7%90%D7%9C-%D7%A1%D7%99%D7%9B%D7%95%D7%9D-%D7%95%D7%AA%D7%95%D7%91%D7%A0%D7%95%D7%AA-%D7%A9%D7%9C-%D7%95%D7%95%D7%A2%D7%93%D7%AA-%D7%9E%D7%95%D7%9E%D7%97%D7%99%D7%9D-2020.pdf

Kotkin, J., 2020. Munk debates – article on website. 9 April 2020 CityLab – article on website. 9 April 2020 CityLab – article on website. 9 April 2020

CityLab – Bloomberg (Accessed 19.12.20) Bloomberg (Accessed 19.12.20) Bloomberg (Accessed 19.12.20 Joel Kotkin: Urban areas have some serious fixing to do. *National Post*, 23 July (Accessed 20.12.20).

Lee, V. J., Ho, M., Kai, C. W., Aguilera, X., Heymann, D., and Wilder-Smith, A., 2020. Epidemic preparedness in urban settings: New challenges and opportunities. *The Lancet Infectious Diseases*, 20 No. 5, 527–529.

Lyons, G., 2020. We need transport planners with superpowers. *Focus*, June. The Chartered Institute of Logistics and Transport.

The Marker: A leading daily newspaper in Hebrew in Israel: July September 2020 special editions, articles and reports which appeared on 24.7.20, 30.7,20, 31.7.20, 4.8.20, 11.8.20, 28.8.20, 2.9.20, 29.9.209.10.20, 15.10.20.

Marshall, B., 2020. Have we reached city limits? Ipsos Mori 12.6.20 Have we reached city limits? | The Ipsos MORI Almanac (ipsos-mori.com) (Accessed 20.12.20).

Mokhatarian, P., 2020. Webinar: The adoption and travel impacts of teleworking: Will it be different this time? *Eno Center for Transportation*, 14 May.

Moreno, C., 2020. The 15 minute city: Fantasy or Reality, Apolitical (Accessed 26.11.20) The-15-Minute-City_-Fantasy-or-Reality_-_-Apolitical.pdf (moreno-web.net) (Accessed 20.12.20).

Natanzon, R., 2020. In Globes 5.6.20.

National Knowledge and Research Center for Emergency Readiness, 2020. *Public Positions and Behavior in Response to the Corona Crisis, Research Project on Crisis Management*. Israel: Minerva Center Haifa University (in Hebrew).

Newman, P. A. O., 2020. Covid, cities and climate: Historical precedents and potential transitions for the new economy. *Urban Science*, 4 No. 3, 32. https://doi.org/10.3390/urbansci4030032.

OECD, 2020a. Building back better: A sustainable, resilient recovery after Covid 19 building back better: A sustainable, resilient recovery after COVID-19 (oecd.org) (Accessed 20.12.20).

OECD, 2020b. Cities policy responses cities policy responses (oecd.org) (Accessed 20.12.20).

Outsite blog, 2020. The best long term + remote work visas for digital nomads. *Outside*, 20 November. The best long term + remote work visas for digital nomads | *Outsite Blog* (Accessed 20.12.20).

Parker, K. et al., 2020. About half of lower-income Americans report household job or wage loss due to COVID-19. *Pew Research Center Social Trends*. About half of lower-income Americans report household job or wage loss due to COVID-19 | pew research center (pewsocialtrends.org) (Accessed 20.12.20).

Pisano, C., 2020. Strategies for post-COVID cities: An insight to Paris en commun and Milano 2020. *Sustainability*, 12 No. 15, 5883. Sustainability | Free Full-Text | Strategies for post-COVID cities: An insight to Paris En Commun and Milano 2020 (mdpi.com) (Accessed 20.12.20).

Planning Administration, Israeli Ministry of Interior, 2020. Evaluation of the impacts of the Corona crisis on urban planning, April (in Hebrew).

Rightmove, 2020. Rental searches for homes with gardens hit record high for the year Rightmove 24.4.20. Rental searches for homes with gardens rise | Property blog (rightmove.co.uk) (Accessed 20.12.20).

Rosen, G., and Razin, E., 2009. The rise of gated communities in Israel: Reflections on changing urban. *Governance in a Neo-Liberal Era Urban Studies*, 46 No. 8, 1702–1722. https://doi.org/10.1177%2F0042098009105508.

Schmidt, M., 2020. Impacts of COVID on urban transport. Preprint: https://doi.org/10.13140/RG.2.2.29901.59362.

Shareable: Sharing Cities: Activating the Urban Commons ISBN: 9780999244005 Publisher: Tides Center/Shareable.

Shenkar, J., 2020. Cities after Coronavirus: How Covid-19 could radically alter urban life. *The Guardian*, 26 March.

Sneader, K., and Singhal, S., 2020. The future is not what it used to be: Thoughts on the shape of the next normal the future is not what it used to be: Thoughts on the shape of the next normal (mckinsey.com) (Accessed 20.12.20).

Tel Aviv-Yafo Municipality's Response to the COVID-19 Pandemic, 2020. [Tel Aviv-Yafo Municipalitys response to the COVID-19 pandemic- May 2020 update.pdf (tel-aviv.gov.il) (Accessed 20.12.20).

Thaler, R., and Sunstein, C., 2017. *Nudge: Improving Decisions About Health, Wealth and Happiness*. New York: Penguin Putnam Inc.

UN Habitat, 2020. Covid-19 response, plan and programmatic framework UN-Habitat's COVID-19 Response plan | UN-Habitat (Accessed 20.12.20).

UNDESA, 2019. World urbanization prospects: The 2018 revision 2018 Revision of world urbanization prospects | Multimedia library – article on website. 9 April 2020 CityLab – article on website. 9 April 2020 CityLab – article on website. 9 April 2020 CityLab – Bloomberg (Accessed 19.12.20) Bloomberg (Accessed 19.12.20) Bloomberg (Accessed 19.12.20United nations department of economic and social affairs (Accessed 20.12.20).

UNESCO, 2020. Learning from city responses to Covid-19 urban solutions: Learning from cities' responses to COVID-19' (unesco.org) (Accessed 20.12.20).

Venn City: Haaretz in south Tel Aviv, a new way of 'neighboring' – article on website. 9 April 2020 CityLab – article on website. 9 April 2020 CityLab – article on website. 9 April 2020 CityLab – Bloomberg (Accessed 19.12.20) Bloomberg (Accessed 19.12.20) Bloomberg (Accessed 19.12.20) and gentrifying – article on website. 9 April 2020 CityLab – article on website. 9 April 2020 CityLab – article on website. 9 April 2020 CityLab – Bloomberg (Accessed 19.12.20) Bloomberg (Accessed 19.12.20) Bloomberg (Accessed 19.12.20) Israel News – article on website. 9 April 2020 CityLab – article on website. 9 April 2020 CityLab – article on website. 9 April 2020 CityLab – Bloomberg (Accessed 19.12.20) Bloomberg (Accessed 19.12.20) Bloomberg (Accessed 19.12.20) Haaretz.com (Accessed 20.12.20).

Venn City. Shapira homepage – article on website. 9 April 2020 CityLab – article on website. 9 April 2020 CityLab – article on website. 9 April 2020

CityLab – Bloomberg (Accessed 19.12.20) Bloomberg (Accessed 19.12.20) Bloomberg (Accessed 19.12.20) Venn City (in Hebrew) (Accessed 20.12.20).

Wintle, T., 2020. COVID-19 and the city: How pandemics could break up our metropolises CGTN Europe 20 July COVID-19 and the city: How pandemics could break up our metropolises – article on website. 9 April 2020 CityLab – article on website. 9 April 2020 CityLab – article on website. 9 April 2020 CityLab – Bloomberg (Accessed 19.12.20) Bloomberg (Accessed 19.12.20) Bloomberg (Accessed 19.12.20) CGTN (Accessed 20.12.20).

Zhang, K., 2020. How urban transport is changing in the age of COVID-19. How urban transport is changing in the age of COVID-19 – article on website. 9 April 2020 CityLab – article on website. 9 April 2020 CityLab – article on website. 9 April 2020 CityLab – Bloomberg (Accessed 19.12.20Bloomberg (Accessed 19.12.20) Bloomberg (Accessed 19.12.20) coronavirus coverage (columbia.edu) (Accessed 20.12.20).

7 Will Covid-19 open the door to more sustainable consumption and lifestyles?

Introduction: household consumption and climate change

In the 1970s, high-income countries, especially Europe, became concerned that imported resources would not be easily available at a reasonable cost for their economic activities and consumption patterns. The growing demand for energy and other resources was framed as a problem of limited supply, not as an environmental problem. OECD and EU initiated policies concerning resource efficiency and productivity, resource accountability, recycling, and waste minimization. The idea of reducing the demand for products, services, or energy uses was outside of the mainstream discourse. Life cycle assessment (LCA) and dynamic system modeling became widely used analytical tools.

During the 1980s, leading scientists raised a new and potentially serious concern: the impact of the growing use of fossil fuels on global climate. The policy measures developed earlier to assure a steady supply of natural resources were now extended to the climate protection policies: increased efficiency, waste minimization, and, in the 2000s, the circular economy. In addition, significant attention was directed to supplementing fossil fuels with renewable sources of electricity, such as wind and solar. Like before, the focus was on technology. Technological solutions are attractive for several reasons: they are linked to modernity and progress; spur innovation and job creation; promise to meet the needs of a growing economy without putting any constraints on dominant lifestyles, institutions, and power relations; and can be accomplished through familiar regulatory approaches.

Environmental organizations also liked that approach because it gave them an opportunity to clearly draw the battle lines: on the one side was the fossil fuel industry and some fuel-intensive industries, as well as industrial agriculture (with its heavy use of toxic chemicals) and on the other side were renewable energy technologies. Such framing of the climate problem fits into the time-honored strategies and tactics of environmental organizations.

The connection between climate change and household consumption and lifestyles has only been made relatively recently, and is still not widely recognized (Dubois et al., 2019). Governments recognized that consumption and lifestyles are problematic but faced great challenges in developing appropriate policies. The prevailing ideological notion of "consumer sovereignty" makes interventions in lifestyles politically difficult. In addition, the widely adopted accounting systems for greenhouse gas (GHG) emissions did not reflect consumption at all. Following the International Panel on Climate Change (IPCC) standards, GHG emissions were counted at the site of production, not consumption. In the global economy, this meant that China became the biggest emitter of GHGs, even though their products were largely purchased and used in Europe and North America (Isenhour and Feng, 2016).

Which household activities are mainly responsible for GHG emissions? Research done in the context of the SCORE! project showed that most emissions stem from transportation, heating, cooling, building of houses, food and beverages, and leisure activities (Tukker et al., 2008; Akenji et al., 2019). The emission intensities are shown to be correlated with income: the higher the income, the higher the amount of GHG emissions. This is true both within countries and between countries (Weber and Matthews, 2008; Wiedenhofer et al., 2017). At the same time, within each income category there is a large spread of emissions. This indicates that there are large opportunities for reducing consumption within income categories. However, it is not clear what the key factors in this wide variation are. Counterintuitively, adoption of pro-environmental attitudes is not correlated with the level of GHG emissions. People who consider themselves green do not emit less than their indifferent neighbors; and because they are often richer and more privileged, they probably emit more.

Historical context of consumption and consumerism

Modernity harnessed science and technology for the enhancement of human wellbeing, which created wealth and leisure through the production of materials- and energy-consuming goods and brought many benefits to mankind. The downside of this progress has been that humanity came to treat nature as a resource and a sink for waste. By the time of the realization that the Earth's life supporting systems are limited, and the limits have become quantifiable, much of the humanity has been entrenched in the culture of consumption and growth. The crucial question is whether the consumer society can continue to be the main pillar of economic and social wellbeing of society, as it has done for the past seven decades, while also respecting the planetary and societal boundaries (Rockström et al., 2009; Raworth, 2017). And if not, as is most likely, can it adapt to the growing demands of sustainable future?

Consumer society is a complex system of technology, culture, institutions, markets, and dominant business models. It is driven by the ideology of neoliberalism and infinite growth. It has evolved through a sophisticated exploitation by the advertising industry of the fundamental human quest for a meaningful life and wellbeing (Skidelsky and Skidelsky, 2012; Sterman, 2014; Speth, 2008; Lorek and Fuchs, 2013; Brown and Vergragt, 2016). Its advent is generally placed in the first two decades of the 20th century in the US. In this period, marketing experts and major think tanks, supported by big business, laid the groundwork for consumer society. The provident remark made in 1929 by Charles Kettering, director of research at General Motors, has been widely quoted by various authors: *"The key to economic prosperity is the organized creation of dissatisfaction . . . . If everybody was satisfied, nobody would want to buy anything"* (Kettering, 1929, cited in Botsman and Rogers, 2010: 35, 250).

The National Association of Manufacturers supported think tanks and various campaigns to create a demand for domestic goods and popularized the term "consumer" while referring to the American people. By the 1930s, both the government and labor unions, in addition to private business, actively supported the consumerist lifestyles of the population as the path to full employment and improved living conditions.

Contemporary historians and sociologists have extensively recounted the corporate strategies then and now to grow consumer demand through aggressive marketing and advertising, tailored to various social, gender, and age groups, including young children (Schor, 1992, 1998, 2004). The end of the World War II is generally considered to be the point when the growth of consumer society took off in the earnest (Cohen, 2004). In the US, the 1944 GI Bill helped returning war veterans to get free college education as well as down payments and government-guaranteed loans for purchasing homes and other goods, and car-dependent sprawling suburbs were created. National output of goods and services doubled between 1946 and 1956, and doubled again by 1970, with private consumption expenditures settling at about two-thirds of the GDP (today it is 70% of GDP). This transition occurred through simultaneous efforts of government, organized labor, and the manufacturing sector. It changed lifestyles of most Americans in profound ways, but it also fostered a cultural shift: consumerism and suburban lifestyle became conflated with such fundamental aspirations as wellbeing, freedom, democracy, and the so-called American Dream.

Research during the past two to three decades has advanced our understanding of consumption in the modern society: what drives it, how it sustains itself, and how the mainstream culture, institutions, and business models shelter it from challenges. It has revealed the link between economic growth, political stability, and consumption, and disclosed the link between consumerism and inequalities. Here, we can only provide a short summary.

Scholars have tried to make a distinction between "needs" and "wants", between what you "really" need to survive and what you want in order to satisfy greed that is supposed to be less necessary. Such a distinction is very hard to make; it depends on place, time, culture, and social class; and it changes all the time. For instance, 30 years ago the internet did not exist; and now a Wi-Fi at home is a basic necessity.

Conceptually, a useful distinction has been made by Manfred Max-Neef: the distinction between needs and satisfiers (Max-Neef, 1991; Guillen-Hanson, 2017). Needs are considered basic and immutable, independent of place, time, and cultural context. Satisfiers, on the other hand, are the means to satisfy our needs: a car, a bike, or a horse for transportation; a dream house or a mobile home for living; a beefsteak or a hotdog for food, etc. Satisfiers are thus highly contingent and determined by cultural context, purchasing power, place, and time. A meaningful policy to address overconsumption thus needs to focus on satisfiers and how to modify them.

Consumption is also intertwined with what people consider to be a good life and what people consider a successful life. A lot of research on human wellbeing has created insights in what constitutes a good life; and the evidence is convincing that once the basic needs are satisfied, wellbeing is less related to material goods and more to relationships, family, successful career, and acceptance by a community of peers. But wants and satisfyers are relative and are socially and culturally constructed. In the 1950s, a 1,000 sq. ft. house with one and half bathroom and a backyard gave a middle-class family a sense of wellbeing. Today, that baseline consists of a 2,200 sq. ft. house with at least three bathrooms.

Furthermore, in the highly competitive society that embraces materialistic values, acceptance by a community of peers becomes linked to the size of the house, quality of the furniture, and the type of family vacation (Layard, 2005). This is often referred to as "keeping up with the Joneses". The fact that our society recognizes success through money – fame leads to higher income and monetary prizes – indirectly cements the relationship between the sense of wellbeing and material goods that can be purchased with that money (Kahneman and Deaton, 2010). Similarly, the clothes and the jewelry you wear and your attributes like handbags, watches, ties, and scarfs, all signal social status. In addition, psychologists have investigated the act of shopping, which is apparently very satisfying to mitigate anxiety and stress, and to create a feeling of fulfillment.

Consumption has also become associated with celebrations such as weddings, baby showers, Christmas, birthdays, and jubilees. Despite the research findings that the act of giving is more important than the actual present, more and more expensive gift-giving has become increasingly important for a successful celebration; most consumers in rich cultures and countries are already so stuffed with consumer goods that they hardly need anything in addition.

Business, industry, and advertisement companies have understood this for a long time and developed many strategies to seduce and encourage buying more of their products, thus boosting growth of business and the economy as a whole. In addition, this "consumption-industrial complex" creates a lot of jobs, which in turn create the purchasing power necessary to keep this cycle going.

Consumption enters the ecological discourse

Calculations show that a switch to 100% renewable energy is necessary but not sufficient to mitigate climate change (Alfredsson et al., 2018). First, the amount of energy (largely fossil fuels) that is necessary to build the renewable infrastructure (wind farms, solar power grids) far exceeds the available carbon budget that would allow us to stay below 1.5°C. Second, bringing the poorest segments of the global population to the level of consumption necessary to satisfy the most basic needs will alone consume the majority of that 1.5°C carbon budget. Clearly, consumption in the rich countries needs to diminish.

One of the problems with the lack of recognition of the role of consumption in GHG emissions was the way the emissions have been counted. Following the lead of the IPCC, inventories around the world – ranging in scale from small towns to countries, regions, and the globe – have looked at direct emissions from individual economic sectors and within geographic boundaries. For example, a traditional inventory might track emissions from heating, cooling, and powering a shopping mall but does not interrogate who uses the mall and for what purpose. It may track emissions from aviation but does not investigate who travels by plane and why.

The introduction of consumption-based emission inventories (CBEI), mentioned in the preceding paragraph, had a profound effect on how we look at consumption and emissions. CBEI tracks emissions from the actual activities of the principal social entities, such as households, businesses, government activities, and other institutions (C40, 2018; Davis & Caldera, 2010). Based on life cycle analysis, the CBEI accounts for both direct use of energy (for instance, heating and electricity) and the embodied energy in materials purchased and used, regardless of where they were manufactured. Using this approach, the state of Oregon, a leader in adopting CBEI, estimated that in 2015 a whopping 80% of all GHG emissions originated in households: the embodied and operational energy of houses and their furnishings, food consumption, leisure travel, car driving, and so on (the other 20% were associated with government operations and businesses) (Oregon, 2018).

The need to consider consumption patterns from a sustainability perspective was first recognized in Agenda 21 at the global summit on "Environment and Development" in Rio de Janeiro in 1992 (United Nations, 1992):

> 4.5. *Special attention should be paid to the demand for natural resources generated by unsustainable consumption and to the efficient use of those resources consistent with the goal of minimizing depletion and reducing pollution. Although consumption patterns are very high in certain parts of the world, the basic consumer needs of a large section of humanity are not being met. This results in excessive demands and unsustainable lifestyles among the richer segments, which place immense stress on the environment.*

A first and widely quoted definition of "sustainable consumption" is from the Oslo Symposium of 1994:

> the use of goods and services that respond to basic needs and bring a better quality of life, while minimizing the use of natural resources, toxic materials and emissions of waste and pollutants over the life cycle, so as not to jeopardize the needs of future generations.
>
> (International Institute for Sustainable Development, 1995)

To quote further:

> Sustainable consumption is an umbrella term that brings together a number of key issues, such as meeting needs, enhancing the quality of life, improving resource efficiency, increasing the use of renewable energy sources, minimizing waste, taking a life cycle perspective and taking into account the equity dimension.
>
> (ibid.)

Twenty-three years after Rio, sustainable consumption was further recognized as one of the UN Sustainable Development Goals (SDGs) (Sustainable Development Goals, 2015). Goal 12 states *"Ensure sustainable consumption and production patterns"*. It consists of 13 subgoals; the first is "*Implement the 10 Year Framework of Programmes on Sustainable Consumption and Production Patterns, all countries taking action, with developed countries taking the lead, taking into account the development and capabilities of developing countries*". This 10YFP contains a number of programs including one on sustainable lifestyles (10YFP).

In addition to the UNO, Pope Francis also recognized consumption as an ethical problem and declared in his 2015 Encyclical Laudato si': "*The pace of consumption . . . has so stretched the planet's capacity that our contemporary lifestyle, unsustainable as it is, can only precipitate catastrophes*" (Laudato Si, 2015). This statement reflects a wider change among faith-based groups recognizing the more dimension of our consumption patterns in relation to environmental issues.

In the preamble of the Paris Agreement on Climate Change (Paris Agreement, 2016) in which nearly all countries agreed on carbon emission targets and on implementing policies to mitigate climate change, it was mentioned: ". . . *Also recognizing that sustainable lifestyles and sustainable patterns of consumption and production, with developed country Parties taking the lead, play an important role in addressing climate change*" (Preamble Paris Agreement, 2016). However, in the agreement itself there was no mention of changing consumption patterns.

The issue of sustainable lifestyles was and is politically fraught with conflict, particularly an inherent conflict with the overriding global goal of economic growth, as measured by GDP and based on increasing consumption to drive the economy. It also conflicts with the widely held neoclassical ideology of "consumer sovereignty", in which a consumer is supposed to be freely expressing his needs and wants (utilities) on the global marketplace. In the still prevalent neoclassical economic theory as widely taught and used by policy makers and think tanks, economic growth and consumer sovereignty are cornerstones of policies that, as we now know, have led to climate change and inequality (Piketty, 2014).

Today, in the context of climate change, there is increasing global recognition that it will not be possible to achieve the goal of remaining below 1.5°C global warming by technological innovation alone (Alfredsson et al., 2018). The climate change and resource management agendas are introducing the critical issues of socio-economic transformations into the discussion. Many proposals for incorporating changes, however, met severe resistance and barriers.

Covid-19 has now removed some of those barriers, enabling discussions that were marginal to become mainstream. Following Covid-19, it is now possible to question economic growth as a measure of societal wellbeing and the dependence on globalized supply chains. Mainstream politicians have recognized that in times of crisis the national government has an essential role to play to support income, provide employment, and provide a safety net and the "essential" public services. It is in this context that we therefore propose that it is opportune to link the societal implications of Covid-19 to the societal implications of unsustainable consumption and lifestyles.

The collapse of the economy during the Covid-19 pandemic has made abundantly clear what are essential products and services, and what not. Obviously, healthcare and healthcare workers, services, and products are essential; what has been exposed during the crisis are the low wages of many health workers and supporting personnel like cleaners and delivery people, as well as the long and vulnerable supply chains for essential products like Covid-19 testing equipment and ventilators for intensive care units. Other consumption habits like eating in restaurants, faraway holidays, and expensive business lunches and dinners proved to be dispensable; however, it is unclear how persistent these changes will be. Parties and celebrations, as well as sports events, are also cancelled; however, there are clear signs that people consider them indispensable.

Reducing the footprint of consumption

Initial approaches to advance sustainable consumption were based on the assumption that there was no basic conflict with economic growth and that society could continue to consume as it wished provided that the energy and materials it used and the wastes it produced were less damaging to the environment. "Greening" consumption by technological advances in production processes and product design was easily adopted in policies and happily accepted by the public. One example is eco(re)design of products: the (re)design of products using less energy and materials for production; reducing energy in use and enabling easy disassembly to advance reuse of parts and recycling; enhancing the technical lifespan of products through durability and increased quality; and taking into account the entire life cycle of the product, from cradle to cradle (McDonough and Braungart, 2002). Part of ecodesign is the use of "green" or "natural" materials. Biodegradable materials reduce potentially the amount of waste. Another example is dematerialization in order to reduce the consumption of resources. Through redesign, products are developed with less materials and less embodied energy.

Since the 1980s, academics, designers, activists, and policy makers have been working on these concepts, stimulated by the increasing waste stream, the perceived shortage of raw materials, and the environmental and social effects of mining virgin materials. They created the "waste hierarchy", which is a hierarchical tool to prioritize strategies for dealing with waste – from prevention through reuse and recycling to incineration and landfill as the least desirable options (DEFRA, 2011). More recently, the Ellen

McArthur Foundation, an industrial think tank, reinvented and popularized this strategy under the catchy title of "circular economy"; it is, on the one hand, a positive development; on the other hand, it is disappointing to notice that industry since the 1980s basically has made little progress beyond the waste hierarchy.

Governments and the business world adopted since the 1980s and 1990s "green consumerism" through regulation, economic incentives, and marketing. Green labeling became a popular tool to incentivize consumers to buy "green" products, at first developed by consumers' and producers' organizations, and later mandated by governments. Examples are energy labels for electrical appliances and labels indicating that products could be recycled. In Europe, the concept of "extensive producer responsibility" was adopted in the 1990s (OECD, 2001). In Germany, the installation of solar panels and other renewable energy appliances was stimulated by feed-in tariffs: basically, the price utilities would pay to consumers who became producers of electricity (prosumers), which was mandated or subsidized by governments.

This technological greening of consumption was undoubtedly beneficial and brought increasing awareness and incremental improvements to reducing emissions and protecting natural resources, but it did not reduce GHG emissions and did not question the fundamental issues of unsustainable consumption. Rather than forcing the responsibility on individual consumers, the issues are the ideology of socioeconomic institutions and the existence of infrastructures that frame choice. The most important is the ideology of perpetual economic growth and the validity of its main indicator, the gross domestic product (GDP). GDP as an indicator of economic wellbeing has been criticized for a long time (Hayden and Wilson, 2017), and many attempts have been made to replace it with a less reductionist and more inclusive concept that comes closer to wellbeing. An important example of such an inclusive concept is the Human development Index (HDI) developed by Amartya Sen and Mahbub ul Haq in 1990 (Human Development Index, 1990)

Perpetual economic growth has been criticized by many leading economists since the 1990s (Daly, 1997) from a variety of perspectives. One is economic growth "does not lift all boats" and is not the best means to address inequality and poverty (Stiglitz, 2013). Another criticism originated from the ecological perspective; contrary to the green growth optimists that we need economic growth to spur innovation and, in that way, address environmental pollution, the more fundamental criticism is that growth by definition cannot respect ecological boundaries. The concept of ecological boundaries is rather new (Rockström et al., 2009) but the fundamental issue that perpetual growth is not possible goes back to Georgescu-Roegen (1971) and the Stockholm Environmental Conference

(1972) and to the Limits of Growth report of the Club of Rome (Meadows et al., 1972).

Directly addressing and criticizing the growth paradigm, the degrowth movement, which started in Southern Europe (D'Alisa et al., 2014), has now spread across the globe and includes academics and activists. Not only it propagates a new economic order and theory that includes the reality of ecological boundaries (Raworth, 2017), but also it makes the connection with alternative economic activities "at the grassroots" that are not guided by profits, competition, and growth (Kawano, 2013). Examples are producer and consumer cooperatives, co-housing initiatives, ecovillages, slow money, slow food, and other non-commercial sharing economy activities. We will later return to these under the heading of lifestyles.

A post-Covid-19 transition to sustainable lifestyles?

Considerable research has been conducted in the past decade on the motivations and the willingness of people to reduce their carbon footprint by changing their daily practices or more fundamental dimensions of their lifestyles. The results generally show that such change is very challenging and the results in terms of carbon footprint are modest. Green consumption, recycling, organic food, and fair trade are widely cited among environmentally conscious consumers but are insignificant in impacts. Environmentalists have the same footprint as non-environmentalists (Csutora, 2012). The highly touted shared or collaborative consumption has been disappointing: car sharing has actually increased traffic and gasoline use, and clothes rental has not affected the culture of consumerism and has, in fact, increased it by opening up the range of options for more clothing. Tool sharing and repair cafes have been more promising in reducing purchases and deemphasizing the culture of consumption, but the impacts on carbon emission are minimal. Eco-communities have demonstrated that lifestyle changes can significantly reduce carbon footprint of a household but, so far, they are rather marginal and inward-centered social arrangements. And so have been the practitioners of voluntary simplicity, minimalism, tiny house living, and other non-mainstream living arrangements. Probably, the most effective are vegetarianism and veganism.

In the EU-funded GLAMURS (Green Lifestyles, Alternative Models and Upscaling Regional Sustainability) project, 14 sustainability initiatives in seven case study regions across Europe have been investigated in order to advance the understanding of drivers of and barriers to sustainable lifestyles, the effects on wellbeing and on members' environmental footprint, and the potential for a widespread transition toward a sustainable society (GLAMURS, 2017). The research covered six lifestyle domains:

work-leisure balance, housing, food consumption, mobility, energy use, and the consumption of manufactured products. Cases include both comprehensive lifestyle change initiatives (e.g., transformations of all aspects of lifestyles) and issue-based initiatives (e.g., transformation of energy sources and end-use patterns of behavior) with potential implications for sustainable lifestyle transformations.

The researchers found that citizens with high levels of overall social engagement, perceived self-efficacy, and environmental identity are more likely to engage in sustainability initiatives. They also found that, in addition to pro-environmental motives, a desire for meaningful social connection drives involvement in sustainability initiatives. Participants in sustainability initiatives also reported a higher level of wellbeing than non-participants. The research also showed that voluntary initiatives even among the highly committed environmentalist can only go so far. In one study, focus groups that probed what practices people were willing to change in their everyday lives showed large reductions in carbon emissions associated with clothing purchases and food consumption (86% and 43%, respectively) but no changes associated with transportation and housing. Overall, the participants in sustainability initiatives had a 16% lower footprint than non-participants. Members of European environmental grassroots initiatives reconcile lower carbon footprints with higher life satisfaction and income increases (Vita et al., 2020). It appears that a larger reduction would require a change in people's conception of what constitutes a good life as well as government pricing policies and infrastructure changes. These and other reports focusing on changing lifestyles toward sustainability illustrate the challenges of social change toward sustainability and the embeddedness of lifestyles in culture and economy.

Restrictions, both government-imposed and driven by health concerns, adopted during the Covid-19 crisis led initially to a sudden drop in emissions (Le Quéré et al., 2020) as economies and travel came to a virtual halt. Some of the changes added to people's sense of wellbeing. People discovered the pleasures of slower, less harried lives. They discovered the joys of local outdoor recreation and short trips to nearby outdoors, which resulted in record sales of camping and athletic equipment. Walks in the woods and parks replaced visits to shopping malls as a form of family recreation. Home cooking as well as backyard and community agriculture became very popular. And many workers who did not have to carry the burden of simultaneously being employees, caregivers, and teachers welcomed the opportunity to work from home.

But as the pandemic fatigue set in, and as gradual opening of the societies began in the spring of 2020, people eagerly returned to various pre-pandemic activities, such as commuting, modest air travel, social gatherings, and attending restaurants and clubs. The resistance to the return of the restrictions in the fall of 2020 has been growing, partly on ideological grounds (as a matter of individual freedom), partly owing to their profound effect on the economy – especially small business – and partly out of the general fatigue, especially impediments of human contacts. Its manifestations include demonstrations, intentional breaking of rules, refusal to wear masks, and politization of the pandemic-related rules of behavior. Conspiracy theories, such QAnon (https://en.wikipedia.org/wiki/QAnon), found a fertile ground among the resistors.

The Covid-19 crisis shows that in a crisis situation, governments can mandate deep lifestyle changes, but only for a limited period of time. When the crisis perseveres, people get tired and want to go back to the situation as before. This finding is a compelling piece of evidence for what has become a common knowledge among climate activists: that talking about a crisis and looming disasters leads quickly to a fatigue and has a very limited power to change behaviors and culture. If we want to see change, the alternatives would have to be more attractive, easy to adopt, and give people a sense of wellbeing. This has been captured in a definition of sustainable lifestyles in a recent UN report:

> A sustainable lifestyle minimizes ecological impacts while enabling a flourishing life for individuals, households, communities, and beyond. It is the product of individual and collective decisions about aspirations and about satisfying needs and adopting practices, which are in turn conditioned, facilitated, and constrained by societal norms, political institutions, public policies, infrastructures, markets, and culture.
>
> (Vergragt et al., 2016)

One of the conditions is creating infrastructures (for energy, transportation, food provision, and housing) that enable people to live sustainably while improving their quality of life. For city planning, this means, for instance, creating proximity of living, working, shopping, and recreation, so that both car transportation and public transportation are unnecessary for most daily tasks, as elaborated in Chapter 6.

For civil society organizations, the challenge might be to deepen the roots for the newly discovered enjoyable practices of localized recreation, community agriculture, home cooking, and less international travel; to back up the newly discovered connections with nature and with a slower live; and to help people realize that with less shopping and other costly activities they can flourish on a smaller budget, and that what they missed above all during the lockdown were community and close family contacts.

The Covid-19 crisis can be considered an unplanned experiment in a degrowth economy. Forced degrowth is never an attractive option and this is what the experiment confirmed; but lessons may be learned when it becomes necessary to manage a sustainable degrowth process (Victor, 2019). The pandemic laid bare the massive social inequalities: between essential blue-collar workers, such as meat packers, transportation workers, caterers, or Amazon warehouse employees, who had to show up at work regardless of the risks, and white-collar workers, who could work from home. It also revealed the vulnerability of hourly service workers in restaurants, entertainment, travel, and hospitality industry. A growing interest in unionization is indicative of these jarringly exposed social inequities.

One of the lessons from the crisis is that government has critical roles to play in the modern society: to provide adequate financial support for unemployed workers and small businesses; to regulate public spaces and large cultural, political, and sports events; to set an example and sometimes take unpopular short-term decisions for long-term societal benefit; to communicate difficult reality and unwanted necessities to the public. The recent growing interests in the concept of universal basic income (Standing, 2020) and services (Coote and Percy, 2020) reflect the changing attitudes toward the role of government. It will be interesting to see how the neoliberal ideology, which has for decades dominated the US and European politics and policy, will fare in the years following the pandemic. If the idea of markets as the organizing principle of social life were to decline, it will no doubt affect the human relations on individual and community levels toward more emphasis on collective good.

The environmental gains from changes in behavior during Covid-19 may not sufficiently help to achieve climate change mitigation levels without deeper structural and cultural changes in society. Although some positive trends may persist (less air travels for holidays and business; more work-from-home; more recreation locally), this is far from enough to achieve the objectives of the Paris accords, which call for a yearly reduction of CO_2 emissions by at least 5% for industrialized countries. One cannot expect such changes from lifestyle changes by individual consumers alone without strong financial incentives, regulation, and infrastructural changes mandated by governments and enabled by business. We may hope that the trillions of taxpayer dollars that are already being spent on mitigating the Covid-19 crisis would be spent on building this sustainable infrastructure, which at the same time would create a lot of employment (see also Chapter 8). At this point, we should make the objectives and policies of the Green New Deal conflate with the objectives of corona mitigation.

References

10YFP. www.unenvironment.org/explore-topics/resource-efficiency/what-we-do/ one-planet-network/10yfp-10-year-framework-programmes (Accessed 17.12.20).

Akenji, L. et al., 2019. 1.5-degree lifestyles: Targets and options for reducing lifestyle carbon footprints. Technical Report. Institute for global environmental strategies, Aalto university, and D-mat ltd. 2019. Institute for global environmental strategies, Hayama, Japan. 15_Degree_Lifestyles_MainReport.pdf (iges.or.jp) (Accessed 17.12.20).

Alfredsson, E., Bengtsson, M., Szejnwald Brown, H., Isenhour, C., Lorek, S., Stevis, D., and Vergragt, P., 2018. Why achieving the Paris Agreement requires reduced overall consumption and production. *Sustainability: Science, Practice and Policy*, 14 No. 1, 1–5. DOI: 10.1080/15487733.2018.1458815.

Botsman, R., and Rogers, R., 2010. *What's Mine Is Yours: The Rise of Collaborative Consumption*. New York: Harper Collins.

Brown, H. S., and Vergragt, P. J., 2016. From consumerism to wellbeing: Toward a cultural transition? *Journal of Cleaner Production*, 132, 308–317. http://dx.doi. org/10.1016/j.jclepro.2015.04.107.

C40, 2018. www.c40.org/researches/consumption-based-emissions (Accessed 17.12.20).

Cohen, L., 2004. *Consumer Republic: The Politics of Mass Consumption in Postwar America*. New York: Vintage Books.

Coote, A., and Percy, A., 2020. *The Case for Universal Basic Services*. Chichester: John Wiley & Sons.

Csutora, M., 2012. One more awareness gap? The behaviour – article on website. 9 April 2020 CityLab – article on website. 9 April 2020 CityLab – article on website. 9 April 2020 CityLab – Bloomberg (Accessed 19.12.20) Bloomberg (Accessed 19.12.20) Bloomberg (Accessed 19.12.20) Impact gap problem. *Journal of Consumer Policy*, 35, 145–163. https://doi.org/10.1007/s10603-012-9187-8.

D'Alisa, G., Demaria, F., and Kallis, G. (Eds.), 2014. *Degrowth: A Vocabulary for a New Paradigm*. Abington, PA: Routledge-Earthscan.

Daly, H., 1997. *Beyond Growth: The Economics of Sustainable Development*. Boston, MA: Beacon Press.

Davis, S. J., and Caldera, K., 2010. Consumption-based accounting of CO2 emissions www.ncbi.nlm.nih.gov/pmc/articles/PMC2851800/ (Accessed 17.12.20).

DEFRA, 2011. Guidance on applying the waste hierarchy. www.gov.uk/government/ publications/guidance-on-applying-the-waste-hierarchy (Accessed 17.12.20).

Dubois, G. et al., 2019. It starts at home? Climate policies targeting household consumption and behavioral decisions are key to low-carbon futures. *Energy Research & Social Science*, 52, 144–158. https://doi.org/10.1016/j. erss.2019.02.001.

Ellen McArthur Foundation. www.ellenmacarthurfoundation.org/ (Accessed 17.12.20).

Georgescu-Roegen, N., 1971. *The Entropy Law and the Economic Process*. Cambridge, MA: Harvard University Press.

GLAMURS, 2017. file:///C:/Users/pverg/OneDrive/Documents/SCORAI%20 global/supporting%20materials/Final-Report-GLAMURS_baja-res.pdf p. 16 (Accessed 17.12.20).

Guillen-Hanson, G., 2017. Identifying the strategic conditions to develop and strengthen sustainable social innovations as enablers of sustainable living through participatory processes. In: *The 21st Century Consumer: Vulnerable, Responsible, Transparent?* Bala, C., and Schuldzinski, W., eds. Verbraucher Zentrale. NRW. Germany E-Book_21st_Century_ICCR_2016_ (4).pdf (Accessed 17.12.20).

Hayden, A., and Wilson, J., 2017. Beyond-GDP indicators; Changing the economic narrative for a post-consumer society? In: *The Coming of Post-Consumer Society: Theoretical Advances and Policy Implications*, Cohen, M., Szejnwald Brown, II., and Vergragt, P. J., eds. Routledge. www.routledge.com/Social-Change-and-the-Coming-of-Post-consumer-Society-Theoretical-Advances/Cohen-Brown-Vergragt/p/book/9781138642058.

Human Development Index, 1990. http://hdr.undp.org/en/content/human-development-index-hdi (Accessed 17.12.20).

International Institute for Sustainable Development, 1995. The imperative of sustainable production and consumption. Oslo roundtable on sustainable production and consumption, 26 September 2020. https://enb.iisd.org/consume/oslo004.html (Accessed 17.12.20).

Isenhour, C., and Feng, K., 2016. Decoupling and displaced emissions: On Swedish consumers, Chinese producers and policy to address the climate impact of consumption. *Journal of Cleaner Production*, 134, 320–329. http://dx.doi.org/10.1016/j.jclepro.2014.12.037.

Kahneman, D., and Deaton, A., 2010. High income improves evaluation of life but not emotional well-being. *Proceedings of the National Academy of Sciences*, 107, 16489–16493. https://doi.org/10.1073/pnas.1011492107.

Kawano, E., 2013. Social solidarity economy: Toward convergence across continental divides. www.unrisd.org/thinkpiece-kawano (Accessed 17.12.20).

Kettering, C., 1929. Keep the consumer dissatisfied. *Nation's Business*, 17, 30–31, 79. Keep the consumer dissatisfied, Charles F. Kettering (1929) (wwnorton.com) (Accessed 19.12.20).

Laudato Si. www.vatican.va/content/francesco/en/encyclicals/documents/papa-francesco_20150524_enciclica-laudato-si.html (Accessed 19.12.20).

Layard, R., 2005. *Happiness: Lessons from a New Science*. New York: Penguin Press.

LeQuéré, C., Jackson, R. B., Jones, M. W., Smith, A. J. P., Abernethy, S., Andrew, R. M. et al., 2020. Temporary reduction in daily global CO_2 emissions during the COVID-19 forced confinement. *Nature Climate Change*, 1–7. https://doi.org/10.1038/s41558-020-0797-x.

Lorek, S., and Fuchs, D., 2013. Strong sustainable consumption governance – article on website. 9 April 2020 CityLab – article on website. 9 April 2020 CityLab – article on website. 9 April 2020 CityLab – Bloomberg (Accessed 19.12.20) Bloomberg (Accessed 19.12.20) Bloomberg (Accessed 19.12.20) Preconditions for a degrowth path? *Journal of Cleaner Production*, 38, 36–43. https://doi.org/10.1016/j.jclepro.2011.08.008.

Max-Neef, M. A., 1991. *Human Scale Development: Conception, Application, and Further Reflections*. New York and London: The Apex Press.

McDonough, W., and Braungart, M., 2002. *Cradle to Cradle: Remaking the Way We Make Things*. ISBN: 9780865475878.

Meadows, D. H., Meadows, D. L., Randers, J., and Behrens III, W. W., 1972. *The Limits to Growth*. New York: Universe Books.

OECD, 2001. *Extended Producer Responsibility: A Guidance Manual for Governments*. Paris, France: OECD iLibrary | Extended Producer Responsibility: A Guidance Manual for Governments (oecd-ilibrary.org) (Accessed 19.12.20).

Oregon, 2018. State of Oregon: AQ programs – article on website. 9 April 2020 CityLab – article on website. 9 April 2020 CityLab – article on website. 9 April 2020 CityLab – Bloomberg (Accessed 19.12.20) Bloomberg (Accessed 19.12.20) Bloomberg (Accessed 19.12.20) Oregon greenhouse gas emissions (Accessed 19.12.20).

Paris Agreement, 2016. Adoption of the Paris agreement – article on website. 9 April 2020 CityLab – article on website. 9 April 2020 CityLab – article on website. 9 April 2020 CityLab – Bloomberg (Accessed 19.12.20Bloomberg (Accessed 19.12.20) Bloomberg (Accessed 19.12.20) Paris agreement text English (unfccc. int) (Accessed 19.12.20).

Piketty, T., 2014. Capital in the twenty-first century. www.hup.harvard.edu/catalog. php?isbn=9780674430006.

Preamble Paris Agreement, 2016. www.lewik.org/term/11844/preamble-paris-agreement-climatic-change/ (Accessed 19.12.20).

Raworth, K., 2017. *Doughnut Economics: Seven Ways to Think Like a 21st-Century Economist*. VT: Chelsea Green Publishing.

Rockström, J., Steffen, W., Noone, K., Persson, Å., Chapin III, F. S., Lambin, E. F., Lenton, T. M., Scheffer, M., Folke, C., Schellnhuber, H. J., Nykvist, B., de Wit, C. A., Hughes, T., van der Leeuw, S., RodHe, H., Sörlin, S., Snyder, P. K., Costanza, R., Svedin, U., FalkenMark, M., KarLberg, L., Corell, R. W., Fabry, V. J., Hansen, J., Walker, B., Liverman, D., Richardson, K., Crutzen, P., and Foley, J. A., 2009. A safe operating space for humanity. *Nature*, 461, 472–475. DOI: 10.1038/461472a.

Schor, J. B., 1992. *The Overworked American: The Unexpected Decline of Leisure*. New York: Basic Book.

Schor, J. B., 1998. *The Overspent American: Why We Want What We Don't Need*. New York: Harper Perennial.

Schor, J. B., 2004. *Born to Buy*. New York: Scribner.

Skidelsky, R., and Skidelsky, E., 2012. *How Much Is Enough: Money and the Good Life*. New York: Other Press.

Speth, J. G., 2008. *The Bridge at the Edge of the World: Capitalism, the Environment, and Crossing from Crisis to Sustainability*. New Haven and London: Caravan Book.

Standing, G., 2020. *Battling Eight Giants: Basic Income Now*. London and New York: IB Tauris.

Sterman, J., 2014. Cultural change to sustainable consumption: A dynamic system perspective. Presentation given for SCORAI colloquium on sustainable consumption and social change, 4 September. Tellus Institute, Boston. Life video. www.youtube.com/watch?v=c891j7lnaYk&feature=youtu.be (Accessed 19.12.20).

Stiglitz, J. E., 2013. *The Price of Inequality: How Today's Divided Society Endangers Our Future.* New York: W. W. Norton Company.

Stockholm Conference, 1972. https://sustainabledevelopment.un.org/milestones/humanenvironment (Accessed 19.12.20).

Sustainable Development Goals, 2015. Transforming our world: The 2030 agenda for sustainable development. https://sustainabledevelopment.un.org/content/documents/21252030%20Agenda%20for%20Sustainable%20Development%20web.pdf (Accessed 19.12.20).

Tukker, A., Emmert, S., Charter, M., Vezzoli, C., Sto, E., Munch Andersen, M., Geerken, T. et al., 2008. Fostering change to sustainable consumption and production: An evidence-based view. *Journal of Cleaner Production,* 16 No. 11, 1218–1225. https://doi.org/10.1016/j.jclepro.2007.08.015.

United Nations, 1992. Agenda 21. https://sustainabledevelopment.un.org/content/documents/Agenda21.pdf (Accessed 26.09.20).

Vergragt, P. J. et al., 2016. Fostering and communicating sustainable lifestyles: Principles and emerging practices. United nations environment programme – article on website. 9 April 2020 CityLab – article on website. 9 April 2020 CityLab – Bloomberg (Accessed 19.12.20Bloomberg (Accessed 19.12.20Sustainable lifestyles, cities and industry branch (UN Environment), p. 6. www.oneearthweb.org/uploads/2/1/3/3/21333498/un_communicating_sust_lifestyles_summary_lores_2016.pdf (Accessed 19.12.20).

Victor, P., 2019. *Managing Without Growth: Slower by Design, Not Disaster,* 2nd edition. Cheltenham: Edward Elgar.

Vita, G., Ivanova, D., Dumitru, A., García-Mira, R., Carrus, G., Stadler, K., Krause, K., Wood, R., and Hertwich, E. G., 2020. Happier with less? Members of European environmental grassroots initiatives reconcile lower carbon footprints with higher life satisfaction and income increases. *Energy Research & Social Science,* 60, 101329. https://doi.org/10.1016/j.erss.2019.101329.

Weber, C. L., and Matthews, H. S., 2008. Quantifying the global and distributional aspects of American household carbon footprint. *Ecological Economics,* 66 No. 2–3, 379–391.

Wiedenhofer, D., Guan, D., Liu, Z. et al., 2017. Unequal household carbon footprints in China. *Nature Climate Change,* 7, 75–80. https://doi.org/10.1038/nclimate3165.

8 Policy response to Covid-19

Green recovery or a different pathway?

Setting the scene

The outbreak of Covid-19 in Wuhan and the subsequent lockdown resulted in a very hard hit to the Chinese economy in the first quarter of 2020. However, the speed and force of China's domestic response mean that the country now has few domestic Covid-19 cases and that, so far, the economy is in a V-shaped recovery (Rothman, 2020a,b). On a quarterly basis, the first quarter was down 6.8%, the second quarter was up 3.2%, and the third quarter was up 4.9% (National Bureau of Statistics, 2020a). China has so far taken measured stimulus measures to boost the economy and support the recovery, through increasing spending in certain sectors and taking a range of measures to cushion the impact on firms and households.

Despite the economic challenges and turmoil China faces, the prevailing policy direction remains toward continued and deeper opening up. This was clearly stated in the government's Work Report delivered in May 2020 and has been reiterated in policies and speeches (Li, 2020). The central framing now being used by government as it seeks to adapt to both internal and external challenges brought by Covid-19 and trade war describes China's economy in terms of a "dual circulation" model (Zhong, 2020), part of which aims to stimulate domestic consumption and to explore new spaces for international trade and cooperation, among Belt and Road countries, for instance.

While China faces a turbulent economic and geopolitical landscape and is eager to boost consumption, is the environment still a priority in its political and economic agenda (De Boer and Jiang, 2020)? Green and low-carbon development cannot be realized without changing unsustainable lifestyles and consumption around the world. Covid-19 made a sudden halt on a wide range of production and consumption activities and resulted in an immediate reduction of emissions. However, many have warned the rebound of environmental pressure from economic recovery and retaliatory consumption in the aftermath of coronavirus. Any policy efforts aiming to shift to greener development path must understand well the importance of consumption and lifestyles in green recovery.

This chapter discusses the "greenness" of China's recovery policy and measures, particularly the implications on sustainable transition of lifestyles and consumptions in China. It would be impossible to discuss state policy responses without defining the "state" (Stevis and Bruyninckx, 2006). In studies on the roles of the state in environmental governance of both domestic and global environmental issues, in relation with the market and civil society, on various levels, scales, and issues, the term of state requires a specification first. We define "environmental state" in this chapter as: the institution of central government that is embedded in the environmental politics, which include the top leaders, legislators, ministries, and departments directly participating in national environmental policy making. We are well aware that the environmental decisions made in Beijing are outcomes of power struggles and interests and competitions among organizations at levels from international to local. We argue that the environmental state as defined here the "vantage point" to observe the dynamics of the transformation of the Chinese environmental governance. At the end, the features of China's recovery policy are better understood taking the US Recovery Deal and the European Green Deal as references.

Comparison of greenness of economic recovery packages of the US, EU, and China

Under the pandemic, global greenhouse gas emissions plummeted 17% in April 2020, but emissions rebounded surprisingly quickly as economic activity picked up. Economists believed that economic recovery measures could lead to a rebound in carbon emissions unless they stimulate the flow of funds to clean energy and transportation infrastructure, clean energy research and development, and other green investments. The International Energy Agency (IEA, 2020) has called on governments to put the development, deployment, and integration of clean energy technologies at the center of their economic recovery plans. The IEA and other experts believed that, if done well and rebuilt better, they can create new jobs and even begin to address long-standing environmental imbalances.

In response to the economic impact of Covid-19, the United States, Europe, and China have introduced economic stimulus packages to promote a "green recovery". According to the Rhodium Group (2020) assessment, the United States leads in discretionary stimulus spending in absolute terms ($2.44 trillion) and as a share of GDP (11.4%). The EU has passed or announced stimulus packages, including the most recent $1.36 trillion in EU funds (10.4% of the entire EU GDP). Stimulus packages for China and India are significantly smaller, at $521 billion (3.7% of GDP) and $35 billion (1.2% of GDP), respectively.

To be more specific, in December 2019 and March 2020, the European Commission published the European Green Deal (EU, 2020a) and the first draft European Climate Law (EU, 2020b), which set out a clear "carbon neutral" goal of zero emissions by 2050, and outlined a clear roadmap and policy framework for action to achieve this green ambition. Following the adoption of EU's multi-annual financial framework for 2021–2027 of €1.074 [in 2018 prices] trillion and the Next Generation EU, the temporary recovery instrument created to fuel Europe's recovery from the coronavirus crisis, the package of a total of €1.8 trillion [in 2018 prices] will be the largest package ever financed through the EU budget. The package will also help rebuild a post-Covid-19 Europe, which will be greener, more digital, more resilient, and better fit for the current and forthcoming challenges (EU, 2020b). The European Commission has released two strategic documents: the EU Hydrogen Strategy and the EU Energy System Integration Strategy (EU, 2020c). The two strategy documents complement each other and plan to invest hundreds of billions of euros in the hydrogen energy industry over the next 10 years, closely integrated with the European Green Agreement and the next generation of EU recovery plans, with the goal of achieving climate neutrality by 2050 and providing a boost to the economy.

The U.S. Congress (2020) passed the Coronary Assistance, Relief, and Economic Security (CARES) Act on March 27, 2020, a more than $2 trillion economic relief package that fulfills the Trump administration's promise to protect the people. The bill's primary purpose was to provide direct financial assistance to American workers, families, and small businesses. It is thus clear that climate change is barely addressed in the economic stimulus package announced by the US government in March 2020. For example, the plan includes loans and subsidies to airlines, but says nothing about the requirement for airlines to reduce emissions. Not only that, but the US will continue to support the oil and gas industry. However, no matter what the US government does, the general trend toward "green economic recovery" through international cooperation is unstoppable. Recently, the Chinese People's Association for Friendship with Foreign Countries and the state of California co-hosted the "Green Recovery for Environment, Climate, and Post-Covid-19" conference. Both sides said that they regarded climate change as their top priority in international affairs and would continue to make their voices heard on the world stage in the future together with their partners to promote positive progress in the global response to climate change (Ministry of Ecology and Environment, 2020).

Consumption in China

China's transition around the early 1980s from a centrally planned to a market-led economy within three decades was one of the most dramatic social changes of the 20th century – and certainly one with far-reaching consequences (Morrissey, 2011). A consequence of this rapid transformation has been the growing role of consumers as a (potentially) highly powerful segment of contemporary Chinese society. Being driven and shaped by globalization, economic growth, political modernization, emerging middle classes, industrialization/urbanization, advances in information and communication technologies, and challenges of sustainability, the past four decades have witnessed dramatic changes in the role of consumption in China. In 2019, the per capita disposable income of the country's residents reached 30,733 yuan, a real increase of 4.4 times over 2000 and an average annual real increase of 9.2%. The nation's per capita consumption expenditure in 2019 was 21,559 yuan, increased by 78.9% from 2012. From 2012 to 2019, the per capita consumption expenditure of urban residents increased by 7.3% per annum; per capita consumption expenditure of rural residents increased by 10.4% per annum (Fang, 2020). The annual contribution of final consumption expenditure in 2019 to GDP growth was 57.8% (National Bureau of Statistics, 2020b). As a major consumer society, the volume, structure, and features of what is being consumed by the Chinese have profound impacts in multiple ways. The cultural shift from communism to consumerism challenges the sustainability of the country and, because of its size, also the world as a whole (Zhang et al., 2018).

Before Covid-19, apart from increasing scale, the following trends of consumption had been observed: premiumization (i.e., competition through offering higher-quality items that consumers value) (Macer, 2015), green consumption, and internet-based explosive development of retail. A United Nations Environment Programme (UNEP) report, based on surveys on consumers in ten Chinese cities, shows that more than 70% of the respondents recognized the environmental consequences of consumption behaviors, and only 8.8% of the respondents denied such connections (Li et al., 2017). Although 80% of Chinese consumers expressed their willingness to pay more for brands with a commitment to sustainability, which is higher than the global average of 66% (National Bureau of Statistics, 2018), the increase in the consumer confidence index from 2015 to 2017 is an indication of absolute growth in total consumption (Nielsen, 2018). Recent years have also seen increasing use of the internet and mobile communication when buying consumer goods and services in China. As of March 2020, China had 904 million "netizens", up by 75.08 million from the end of 2018, and its internet penetration had reached 64.5%, up by 4.9 percentage points over the end of 2018. Up to March 2020, the user size of online shopping was 710 million or 78.6% of China's total netizen population, up by 100 million over the end of 2018; the number of mobile shopping users had reached 707 million, up by 116 million from the end of 2018, taking up 78.9% of mobile internet users (China Internet Network Information Center (CNNIC), 2020).

Soon after China was back to normal life from Covid-19, the power of consumption was first tested during the first week of October 2020, also known as "golden week", which was celebrated in China around the national day and the traditional Moon festival, providing the best evidence of retaliatory consumption. While international travel was considered unsafe, "domestic travel" related searches rose by 41.49% year-on-year, far more than the same period last year (TechWeb, 2020). Within these 8 days' holiday, a total of 637 million domestic tourists traveled within the country, generating domestic tourism revenue of 466.56 billion yuan. The national average daily passenger traffic on railways, roads, waterways, and civil aviation was 62.115 million. Passenger traffic on national railways remained above 10 million passengers for 8 consecutive days, with the average daily passenger volume being about 90% of that of the same period last year.

This "golden week" was followed with even more enthusiasm for consumption that could be observed during the 3rd China International Import Expo (CIIE) held during November 5–10, 2020, in Shanghai and then the shopping frenzy called "double 11" in China and beyond. The CIIE 2020 gave firm support to trade liberalization and economic globalization and actively opened the Chinese market to the world. Within 6 days, CIIE attracted more than 2,600 enterprises around the world, and the total intention of economic and trade cooperation reached $72.62 billion, a 2.1% year-on-year increase (CIIE, 2020). "Double 11", also known as "Singles' Day", originated as a shopping festival first initiated by Alibaba in 2009 and joined quickly by other e-commerce platforms, and has transformed itself from the celebration of being single to a nationwide and all-sites-wide shopping fever. Double 11 sees records broken every year, and customers waiting eagerly for the clock to strike at midnight on November 11. This year's shopping festival marks the 12th Double 11. Alibaba was reported to nearly double last year's record with over $74 billion in total sales from November 1 to November 11. If including all the e-commerce platforms, the total transaction value reached nearly a trillion dollars, indicating a further consumption recovery since Covid-19. "Double 11" is therefore being regarded as a barometer of Chinese consumption and economic resilience this year.

China's pre-Covid-19 policy on consumption

China's policy response to the problems associated with unsustainable consumption can be traced back to China's Agenda 21, which was issued in 1994 by the State Council, the highest administration level of the government. However, the promotion of sustainable consumption remained marginal afterward, until recently. Since 2011, the government turned its attention to more sustainable consumption, emphasizing "green consumption", not less consumption. "Green" manifests itself primarily as reducing packaging waste, more energy-efficient appliances, and safer and healthier consumer goods while citizens are encouraged to consume and thus maintain the national economic growth rate at a high level.

At the national level, in March 2011, the 12th Five-Year Plan included "Green Development, Building a Resource-Conserving and Environment-Friendly Society" and "Vigorously Developing the Circular Economy" (Chapter 3) and "Promoting Green Consumption Patterns" and "Incorporating Green Lifestyles and Consumption Patterns" (Section 3). By 2012, sustainable consumption was increasingly framed through the concept of ecological civilization, as proposed officially in the report of the 18th National Congress of the Communist Party. In the meanwhile, the government-driven structural reform since 2015 has been oriented to the supply side as a strategic response to the changing demands in the domestic market for safer, healthier, and environmentally friendly goods and services, and as an opportunity to optimize the supply side. In September 2018, the General Office of the State Council issued "Perfecting the institutions and mechanism for promoting consumption implementation plan (2018–2020)", which aimed to unify green product standards, promote green circulation development, and promote the concept of green consumption in society.

Although the signal for sustainable consumption is strong from the central government, so far, consumption policy measures by different ministries have been fragmented. Ministries tended to target different groups relevant to their own goals. For instance, the Department of Environment and Comprehensive Utilization of Resources of the NDRC and the China Energy Conservation Investment Corporation (CECIC) issued in 2014 the "Guide to Energy Conservation Behavior of the Public", which for the first time is in the form of a guide to direct the public's green consumption behavior, with different guidelines for energy conservation and green consumption behavior for government civil servants and employees of public institutions, urban residents, hotels, and restaurants. The Ministry of Environmental Protection issued in 2015 the "Implementation Opinions on Accelerating the Greening of Lifestyles", focusing on the gradual shift of social lifestyles to green consumption and low-carbon conservation. Ten departments, including the NDRC, the Central Publicity Department, and the Ministry of Science and Technology, issued in 2016 the "Guiding Opinions on the Promotion of Green Consumption". In responding to the rapidly increasing waste from the logistics and express delivery, ten ministries jointly issued in 2017 "Opinions Regarding Greening Packaging from Express Delivery", aiming to reduce, recycle, and green the packaging practices and replace half of the packaging material with disposable materials by 2020. However, there are rarely comprehensive evaluation of the effectiveness and effects of these policies, apart from case-based researches. It remains unclear what data and information need be collected by whom and how they should be analyzed and shared.

Post-Covid-19 policy response

To facilitate the post-Covid-19 recovery from reduced consumption, China has successively issued policies aimed at expanding consumption, greening production and services, and strengthening of environmental protection standards. At the national level, in March 2020, the General Office of the Communist Party of China Central Committee and the General Office of the State Council issued the "Guiding Opinions on Building a Modern Environmental Governance System", which emphasizes the mobilization of all kinds of actors and application of policy instruments like green finance.

At the level of ministries and commissions, in February 2020, with the consent of the State Council, the NDRC, the Central Publicity Department, the Ministry of Education, and other 23 departments jointly issued the "Implementation Opinions on Promoting Consumption Expansion and Quality Improvement and Accelerating the Formation of a Strong Domestic Market", combining green consumption with intelligent electronic equipment and other products, and vigorously promoting "smart radio and television", recycling of motor vehicles and electronic products, and promoting the development of new energy vehicles and other green products. In order to further implement specific work, in March 2020, the NDRC and the Ministry of Justice issued the "Opinions on Accelerating the Establishment of a Regulatory and Policy System for Green Production and Consumption", which aims to further improve regulations, standards, and policies related to green production and consumption by 2025 through the promotion of green design, circular economy, green development of agriculture, green development of the service industry, and expansion of green product consumption. The government has also begun to establish an institutional framework of incentives and restrictions to promote green production and consumption.

Local governments have joined forces with Alibaba and other payment platforms to issue electronic consumer vouchers to boost consumer confidence. The issuance of vouchers through Alipay, Meituan, WeChat, and other platforms can cover the widest range of people. As of May 9, 2020, according to incomplete media statistics, about 80 municipalities had issued vouchers, especially encouraging offline consumption. The distribution of vouchers was combined with the May Day campaign to stimulate spending. Currently, consumer vouchers mainly cover restaurants and department stores, but also car and sports vouchers and travel vouchers have emerged. Vouchers are usually given out in the form of large gift packs and contain several types of vouchers. In terms of current results, there has been a strong boost to consumption growth and a high level of popular participation. The purchasing power of low-income groups in particular has been enhanced. In addition, government participation in the issuance of consumer vouchers is aimed at promoting the development of the real economy (Ali Research, 2020).

The business and corporate sectors are exploring sustainable production and consumption from different perspectives: retailers such as IKEA, Wal-Mart, RT-Mart, and City Shop concentrate on sustainable supply chains; Tetra Pak and Zheng Gu and other manufacturers concentrate on the design, production, use, and recycling stages of the product life cycle. In recent years, a large number of local NGOs such as Green Light Year and Green Earth have emerged in China, and these NGOs are playing an increasingly important role in waste recycling and environmental education.

Compared with developed Western economies, consumption in China is still at an early stage of development. Recognizing the critical goal of building a consumption-driven economy, a systemic transformation of consumption toward more sustainable one has become an important point in the sustainability agenda for China. It has been predicted that by 2030 China will become the world's largest consumer, with an urban middle class of more than 500 million people, and will be the main driver of economic growth in the future. At the same time, with the release of the consumption potential of the huge rural population, consumption, like production activities, will become the main form of interaction between China's population, resources, and environment. China needs to take the fourth wave of the environment movement (Krupp, 2018) as an opportunity to systematically promote the innovation of mindware, hardware, and software.

Implications for future

Mindware: At the national level, innovation in governance concepts cannot follow the same thinking and tools used to govern industry. It needs to be fully recognized that unsustainable consumption and production practices are the social roots of ecological problems and the two driving forces of sustainable transformation. On the basis of the current socio-technological innovation, various effective communication and interaction mechanisms between the government, businesses, and the public, such as one-to-one, one-to-many, and many-to-many, need to be further explored. There is a need to nurture ecological citizenship at the micro-level, and to activate the innovation drive of consumers, communities, organizations, platforms, media, new media, etc., through new institutional provision. To broaden the social debate on the vision of a sustainable lifestyle, particular attention should be paid to educating young children and students on eco-citizenship, so that they can become the main driving force in transforming consumer behavior.

On the software side: this is reflected in the innovation of governance tools. The aim is to establish and promote the construction of information release platform, to innovate the application of deposit-refund mechanism, to explore the sharing economy modes, and to strengthen the credibility and independence of product certification and labeling systems. Sustainable consumption behavior is a specific type of consumption with social awareness and social responsibility, influenced by three different mechanisms: economic, psychosocial, and historical and socio-technical drivers. In addition to the personal characteristics of the consumer, external social norms and the cost of behavior are closely related to the consumer's own sustainable consumption behavior. Consumption is the process of participating in social processes and gaining a certain social status. The influence of the social environment on consumption decisions cannot be ignored, so consumer information on sustainable consumption is crucial. It is recommended that a sustainable consumption competence center similar to the one in Germany be established in the Ministry of Ecology and Environment to integrate sustainable consumption information into ecological and environmental data.

Transition to sustainable consumption is high on the agenda for the unfolding 14th Five-Year Plan, with focus on food, mobility, clothing, housing, leisure, etc. It is equally important to increase awareness at individual level and to improve provision of green goods and services and infrastructure. In the global economic system, China is an important producer and consumer, and actively participates in the global governance of various environmental flows. Systematic monitoring, assessment, and dissemination of information on sustainable consumption will help to improve China's green image in the international arena and promote international cooperation.

Unlike the economic recovery packages of the US, the green develop-ment of China is embedded in different Chinese policies. Although the 2020 Report on the Work of the Government (Li, 2020) did not emphasize green development specifically, green development is still at the core of economic development policy in China. At the macroeconomic level, quantitative annual GDP growth targets have been abolished, and the focus of the expan-sion of investment is on the "two new and one heavy": new infrastructure, new urbanization, and major projects related to people's livelihoods. The aim is to adopt new, more innovative, and cleaner growth patterns, avoiding the practice of relying on large infrastructure and heavy industry to stimu-late macroeconomic growth figures. It is actively guiding society as a whole to establish the concept of green consumption, such as encouraging the con-sumption of electric vehicles, purchasing green and low-carbon products, and supporting low-carbon consumption by individuals, all of which point the way for China's post-epidemic "green recovery". Apart from encour-aging technological innovation and improving the institutional context for the building of an ecological civilization, green development was also codified in urbanization and regional development. For instance, ecological improvement and environmental protection are at the core of "Develop-ment Outline of the Yangtze River Economic Zone" (Political Bureau of the Communist Party of China, 2016) and "Outline of the Plan for Ecologi-cal Protection and High-Quality Development of the Yellow River Basin" (Political Bureau of the Communist Party of China, 2020).

The common feature of these "green recovery" measures is that most of the measures work on the provisioning system/institutional contexts of the economy, and thus lack clear and direct incentives for changing consumers' behaviors toward more sustainable lifestyles. Many have noticed the vast shifts in economy, energy, and technology that are required for achieving China's ambitious carbon targets, but policy efforts have been overwhelm-ingly focusing on increasing rather than decreasing consumption. Adding to this trend, poverty alleviation through consumption was promoted by the State Council Office (2020) as a way for all sectors of society to help the poor increase their income and escape poverty through the consumption of products and services from poor areas and poor people, and this targeted consumption became an important way for social forces to participate in the fight against poverty in China. "The Consumption Poverty Alleviation Action Plan 2020" by NDRC (2020) and other agencies spelt out further targets and accompanying measures. Up to the end of October 2020, more than 227.665 billion yuan have been spent to benefit the targeted population in 22 less developed provinces. This might indicate an innovative approach to achieve broader sustainable transitions with consumption.

References

Ali Research, 2020. www.aliresearch.com/cn/presentation/presentiondetails (in Chinese) (Accessed 11.11.20).

China International Import Expo (CIIE), 2020. www.ciie.org/zbh/en/news/exhibition/News/20201110/24586.html (in Chinese) (Accessed 15.11.20).

China Internet Network Information Center (CNNIC), 2020. Statistical report on internet development in China. http://cnnic.com.cn/IDR/ (Accessed 16.11.20).

De Boer, D., and Jiang, B. Y., 2020. Is the environment still a priority for China in the post-pandemic era? https://chinadialogue.org.cn/en/author/boya-jiang/ (Accessed 30.09.20).

European Union (EU), 2020a. European green deal. https://ec.europa.eu/info/strategy/priorities-2019-2024/european-green-deal_en (Accessed 11.11.20).

European Union (EU), 2020b. Questions and answers on the adoption of the EU's long-term budget for 2021–2027. https://ec.europa.eu/commission/presscorner/detail/en/qanda_20_2465 (Accessed 22.12.20).

European Union (EU), 2020c. EU strategy on energy system integration. https://ec.europa.eu/energy/topics/energy-system-integration/eu-strategy-energy-system-integration_en (Accessed 22.12.20).

Fang, X. D., 2020. Achievements in building a moderately prosperous society in all respects from the perspective of people's income and expenditure. *People's Daily*, 27 July. http://paper.people.com.cn/rmrb/html/2020-07/27/nw.D110000renmrb_20200727_1-10.htm (in Chinese) (Accessed 12.11.20).

International Energy Agency (IEA), 2020. Sustainable recovery: World energy outlook special report. www.iea.org/reports/sustainable-recovery (Accessed 12.11.20).

Krupp, F., 2018. Welcome to the fourth wave: A new era of environmental progress. www.edf.org/blog/2018/03/21/welcome-fourth-wave-new-era-environmental-progress (Accessed 22.03.18).

Li, K. Q., 2020. Report on the work of the government. delivered at the third session of the 13th national people's congress of the people's republic of China. 22 May. www.gov.cn/zhuanti/2020lhzfgzbg/index.htm (in Chinese) (Accessed 13.09.20).

Li, Y., Zhang, L., and Jin, M., 2017. Report on consumer awareness and behavior change in sustainable consumption. Funded by UNEP's the 10-year framework of programs on sustainable consumption and production patterns. www.ccfa.org.cn/portal/cn/ (Accessed 30.06.18).

Macer, T., 2015. Inspiring innovation. *Research World*, *54*, 20–25.

Ministry of Ecology and Environment, 2020. The state of California and others jointly held a video dialogue on "environment, climate and post-epidemic green recovery". www.mee.gov.cn/ywdt/hjywnews/202009/t20200902_796584.shtml (in Chinese) (Accessed 02.09.20).

Morrissey, L., 2011. Consumerist China: Is it sustainable? *Australian Business Review*. www.theaustralian.com.au/business/business-spectator/consumeristchina-is-it-sustainable/news-story/def1baa20cde834451173cab73085075 (Accessed 31.01.18).

National Bureau of Statistics, 2018. Statistical Communiqué of the People's Republic of China on the 2017 National Economic and Social Development. http://www.stats.gov.cn/tjsj/zxfb/201802/t20180228_1585631.html (in Chinese) (Accessed 30.06.18).

National Bureau of Statistics, 2020a. Preliminary accounting results of GDP for the third quarter of 2020. www.stats.gov.cn/english/PressRelease/202010/t20201021_1795384.html (in Chinese) (Accessed 21.10.20).

National Bureau of Statistics, 2020b. Statistical bulletin on national economic and social development of the people's republic of China 2019. www.stats.gov.cn/tjsj/zxfb/202002/t20200228_1728913.html (in Chinese) (Accessed 12.11.20).

Nielsen, 2018. China's consumer confidence index reached all-time high to 114 points in Q4. www.nielsen.com/cn/en/insights/news/2018/Nielsen-China-Consumer-Confidence-Index-reached-all-time-high-to-114-points-in-Q4-2017.html (Accessed 02.02.18).

Political Bureau of the Communist Party of China, 2016. Development outline of the yangtze river economic zone. www.gov.cn/xinwen/2016-09/12/content_5107501.htm (in Chinese) (Accessed 12.11.20).

Political Bureau of the Communist Party of China, 2020. Outline of the plan for ecological protection and high-quality development of the yellow river basin. http://cppcc.china.com.cn/2020-08/31/content_76655714.htm (in Chinese) (Accessed 12.11.20).

Rhodium Group, 2020. It's not easy being green: Stimulus spending in the world's major economies. https://rhg.com/wp-content/uploads/2020/09/Its-Not-Easy-Being-Green-Stimulus-Spending-in-the-Worlds-Major-Economies.pdf (Accessed 22.09.20).

Rothman, A., 2020a. China's economic resilience. *Sinology*, 16 July. https://matthewsasia.com/perspectives-on-asia/sinology/article-1784/default.fs (Accessed 02.11.20).

Rothman, A., 2020b. Four China trends. *Sinology*, 14 August. https://matthewsasia.com/perspectives-on-asia/sinology/article-1794/default.fs (Accessed 12.11.20).

State Council Office of the People's Republic of China, 2020. Guidance on the deepening of consumption poverty alleviation to help win the battle against poverty. www.gov.cn/zhengce/content/2019-01/14/content_5357723.htm?gs_ws=tsina_636831472823423675 (in Chinese) (Accessed 12.11.20).

Stevis, D., and Bruyninckx, H., 2006. Looking through the state at environmental flows and governance. In: *Governing Environmental Flows: Global Challenges to Social Theory*, Spaargaren, G., Mol, A. P. J., and Buttel, F. H., eds. Cambridge: The MIT Press, 107–136.

TechWeb, 2020. Baidu Search Big Data during the National Day. http://www.techweb.com.cn/news/2020-10-09/2806418.shtml (in Chinese) (Accessed 09.10.20).

U.S. Congress, 2020. The coronavirus aid, relief, and economic security (CARES) act. www.congress.gov/116/bills/hr748/BILLS-116hr748enr.pdf (Accessed 18.11.20).

Zhang, L., Liu, W.L., and Oosterveer, P., 2018. Institutional Changes and Changing Political Consumerism in China. In: The Oxford Handbook of Political Consumerism, Boström, M., Micheletti, M., and Oosterveer, P., eds. https://doi.org/10.1093/oxfordhb/9780190629038.013.26.

Zhong, J. W., 2020. Deeply grasping the essence of speeding up the development of a new development pattern. *Economic Daily*, 19 August. www.xinhuanet.com/fortune/2020-08/20/c_1126390928.htm (in Chinese) (Accessed 12.11.20).

9   Conclusions

Introduction

The main objective of this book was to investigate if and how the Covid-19 crisis has opened doors toward sustainable lifestyles. To set the stage, Chapter 2 discussed various social trends affected by the pandemic, new ones as well as those preceding the Covid-19 pandemic and which have been either magnified or attenuated by the virus. The chapter drew on extensive research in Brazil and in the OECD countries. Chapter 3 proposed scenarios for different future lifestyles based on the effects of Covid-19 upon two key lifestyle dimensions: consumption and socializing. Chapter 4 followed up by investigating the various conceptions of home and reviewing the evolution of that concept over the centuries. Taking the sustainability perspective, the chapter emphasized both the physical and human dimensions of a home.

The next two chapters, Chapters 5 and 6, focused on the city and the lives of its inhabitants in the context of a community and a neighborhood, and the impact of the pandemic on their lives. Much of Chapter 5 on communities drew on the Chinese experience, where the community is defined by a geographic proximity and where the term denotes both the human relations and a form of governance from top-down. The Chinese case highlighted the crucial role of communities in fighting the pandemic. Chapter 6 on cities drew on the Israeli experience, a country where the majority of the population is urban. It concluded that the pandemic put to question the type of urbanization that has in recent years appealed to the millennials, especially very dense cities and intense interpersonal interactions. Chapter 7 tells the Western (mostly the US) history of consumer society and the emergence of the awareness that household consumption is the major contributor to overuse of energy and materials in the world and the unsustainable emissions of greenhouse gases. To conclude, Chapter 8 reviewed government policies relevant to sustainability that have been precipitated by the pandemic, especially the slowing down of the economy. The chapter revealed the contrast between the European Union policies, which used the need for an economic stimulus to advance the green economy through large infrastructural investments, and the Chinese policies, which focused on stimulating economic growth, consumption-based economy, and green and socially responsible consumption.

In this chapter, we ask two questions:

- Will the trends we have identified persist beyond the immediate health and economic impacts of the pandemic?
- Can agents of change embed the favorable trends in structures and institutions so that they become an integral part of lifestyles of the future?

First, a word of caution. Since the book is largely based on data from only three countries – China, Brazil, and Israel – generalizations to other regions and countries of the world may be limited and may require a different approach. Trends and policies have not always been coherent or easily understandable even within these countries. Countries have responded to the crisis in a variety of ways, depending on their political, economic, and social contexts, and there is no single relevant approach to coping with its long-term societal implications. Second, at the time of writing, the pandemic is far from over and we are in the second wave of the virus.

The Covid-19 pandemic affected all major dimensions of life and many were radically redrawn. For many, the year 2020 was either a year to forget or, even worse, a year in which tragically lives were lost or took a downward turn in social mobility and distress. The pandemic revealed and heightened flaws, cracks, and faults in many social and economic systems, intensifying inequalities and disrupting supply chains. It transformed the policy agendas of governments from economic growth to public welfare. The future is not necessarily a dichotomy between "going back to old ways" and "accommodating to a new normal" but may consist of multiple combinations between them, generating different and alternative futures.

Although highly speculative at this stage, discussion of the long-term societal implications of the pandemic is important to inform policy making, market decisions, and civil society activities on how to move toward more sustainable lifestyles, harnessing positive trends and restraining negative ones.

Will changes persist?

The first question to be considered is whether current trends will persist, or whether life will go back to "business as usual" as existed before the pandemic.

The acceleration of digitalization stands out among the changes that are likely to be irreversible. Some sections of the population easily adapted to and optimized their life through digitalization. As digital natives, millennials and the upcoming Z generation welcomed and adopted this trend easily. Other sectors of the population moved up the internet connectivity ladder and transformed their operations from face to face to the internet-mediated world. Disadvantaged groups, older generations, informal workers, and many micro-entrepreneurs were left out and struggled to continue direct face-to-face connectivity. Those sections of society that already had or managed to adopt or improve their digital capabilities during Covid-19 in response to quarantine and isolation imposed by containment are highly unlikely to forego the benefits. They gained advantages of teleworking and telemedicine and accessibility to e-commerce and online entertainment.

In contrast, the activities that did not thrive in the digital sphere, such as personal socialization, leisure and cultural experiences, and remote education for young children and adolescents, will revert to the pre-pandemic modes. The connections between socialization and consumer behavior were presented in alternative scenarios in Chapter 3.

The accelerated digitalization has left behind many groups who could not connect or were not connected to digital services. It will therefore be up to policy makers to ensure that accelerated digitalization is made available to all, including disadvantaged groups and to people living in peripheral or remote locations. It will most likely require framing internet as public utilities, similar to electricity, water, and sewage, to which access is guaranteed in the modern society.

Localization is another trend that seems likely to stay. Containment measures generated an increased awareness of the benefits of obtaining supplies and services locally, generated interest in the quality of life in the immediate surroundings of the home and neighborhood, and strengthened local community connectivity. The benefits of living in attractive, well-served neighborhoods with green spaces and a high level of community attachment and organization paved the way to mutual support and social resilience.

Evidence suggests that mindful consumption has taken root and may become an irreversible trend. Mindful consumption goes beyond green or ethical consumption; it interrogates the need for acquisition of additional consumer goods. It is now possible to discuss a reduction of consumerism both in the contexts of financial hardships and changing priorities. For some, it is a question of adaptation to scarcity; but there is also a change toward de-emphasizing material possessions and moving away from seeking pleasure from immediate consumption. Trends show that those minimally affected by the recession are becoming more prudent about spending and save their earnings at a higher rate. It is especially significant because these are the social groups that have previously engaged in conspicuous consumption and consumerist lifestyles.

From a sustainability perspective, a regrettable trend that emerged during the pandemic is the loss of public support and cooperation in reducing waste and willingness to share and to reuse. The use of disposable materials has increased, mostly for self-protection. Online shopping and delivery services were accompanied by a massive increase in packaging. Just as many countries were making progress toward more sustainable habits by legally banning single-use plastic devices and encouraging reuse and shared equipment and services and companies were adapting products to include reusable or recyclable tools, measures for reducing contamination undid these greener practices.

Separation because of the fear of contamination by others brought a return of preference for the private car. Just as many countries were improving public transport and managing to reduce private car use, the pandemic brought back a preference for avoiding contact and the risk of infection in all forms of shared transport. Indications that this may continue are an increase in driving lessons and the granting of driving licenses. It could be further aggravated by public transportation systems' fare hikes and by reduction in public transit services due to financial strains, and by complex requirements for advanced bookings. On the other hand, a boom in sales and use of bicycles is encouraging.

Passive citizenship

Footages of angry citizens revolting against police brutality in the US and quarantine boycotters defying calls for renewed confinement in Europe suggest an upsurge of activism that easily finds its way into ongoing political polarization. However, those images speak of occasional facts, limited in context, and featured by a vocal minority that hides the far larger silence and abdication of rights and liberties by majorities across the world. Authoritarianism was on the rise during Covid-19 times, so much that it consolidated the reversal in democratic versus non-democratic nations balance. States of siege were declared and remained unchallenged, limitations to freedoms were accepted, extended surveillance and social control remained heavily endorsed, and many voting opportunities, be it in Brazil or France, faced record electoral abstention (except in the heavily polarized US presidential election). On a similar vein, foreigners and even nationals from other locations within the same country were deprived from returning home, saying farewell to sick relatives soon dying, while being seen increasingly as sources of blame and targets for discrimination.

The exercise of civic culture by individuals clearly flattened, leaving streets around the world mostly empty. To many leaders, this juncture resolves the problem of governability creating inducements for continuous weakening of democratic mechanism of control of the state and popular sovereignty on grounds of putting lives first and complying with the public health emergency. On the other hand, hygiene fears coupled with material subsistence fears amid a process of radical downward social mobility are highly likely to keep voters away from activism, stretching the life of this passive citizenship cycle.

The importance of supportive policies

The second question is whether changes will become embedded in institutional and organizational processes and frameworks. This section focuses on "agents of change", without whose mobilization positive trends may be lost.

Governments, organized civil society, and business are critical agents of change, who can consolidate and promote sustainable lifestyles through policies, legislation, regulation, facilitation, and actions that anticipate and promote positive trends and restrain or counteract negative ones.

If policy makers see the Covid-19 crisis as a window of opportunity for major progress toward sustainability, positive directions toward more sustainable lifestyles could be embedded in institutional and structural frameworks, to become an integral part of the context in which people make daily choices.

This section focuses on what governments, policy makers, and change agents could and should do to take advantage of emerging trends toward improving wellbeing and promoting sustainable lifestyles and what actions could be taken to counteract and neutralize adverse consequences. Among the former, we identify digitalization, localization, and mindful consumption. Among the latter, we recognize a setback of reuse and sharing and the return to the car.

• Digitalization

The ability to benefit from the acceleration of digitalization is an essential element in inclusive sustainability. The availability of high-quality digital access would offer equal opportunities for remote working, education, shopping, medical treatment, and entertainment and virtual social connectivity.

Digitalization combined with automation and the integration of artificial intelligence (AI) is increasingly seen as the way to revive the global economy and is at the forefront of "Digital Europe". It would provide better public services not only for the prevention, diagnosis, and treatment of disease, but also to open up new opportunities for business and employment in "digital industries".

International organizations, many governments, and the business world have recognized the crucial importance of accelerating digitalization to reach out to a wider public. The private sector is expanding advanced infrastructures, such as fiber optic networks where profitable. The role of governments is to ensure that high-speed internet connectivity is made available to all, including less privileged groups and to peripheral and remote areas, and that assistance is given to small businesses to transform their activities in production and marketing to an internet-mediated framework.

The pandemic revealed that people in less privileged and in peripheral and remote areas suffered since they did not have access to the internet and could not move up the ladder of digitalization. Public investment to provide all sectors of the population with fast access to the internet, wherever they live, has become an essential basic public service requirement. It is not only the technical infrastructure but also the provision of the internet service at low cost, educating the "internet-illiterate", and capacity building to enable all sectors of the population and businesses to use digital access and services. All possible government services should also be available to all through easily understandable portals.

Digitalization can enable people to live where housing is more affordable, where vacant properties are available, to bring back vitality to towns and villages that have suffered from depopulation and ageing population and enable people to enjoy a higher quality of life.

• Localization

The implication of trends on homes, neighborhoods, and cities points to human dispersion and economic decentralization, on the one hand, and to localization and community connectivity, on the other. The results will depend not only on market responses to changing consumer demands but also on the institutional and informal frameworks, which will need to consider how they can regulate or facilitate changes.

Working at or nearby home will bring life into dormitory suburbs, which used to be emptied during working hours; however, it requires regulating authorities to permit non-residential uses in residential areas and enable buildings to change from office use to residential use. Separation of land use activities into residential zones, industrial zones, and commercial zones was already breaking down pre-Covid-19 as the justification for such separations was no longer valid. Preference for mixed uses will require revision of masterplans and zoning regulations and the redesign of neighborhood circulation patterns.

Stringent planning and building regulations will be needed to counteract a possible trend to building larger homes on greenfield sites, which would have very negative consequences of loss of urbanity and loss of ecological habitats and open landscapes. However, relocation of families to existing small towns and villages that suffered a loss of population seeking employment in cities could be very beneficial. That could be facilitated by the provision of high-quality public and private services, including high-speed digital access and the possibility of physical access for hybrid occasional face-to-face meetings.

Building and zoning regulations were originally enacted to protect the health and safety of the public in and outside their homes and workplaces. The value of some of these strict regulations is no longer relevant and they need to be revised to permit flexibility and adaptation, to add balconies and roof gardens, to enable drone delivery services, and to replace a requirement for car parking spaces by spaces enabling micro-mobility such as biking. They could also include support for net zero energy insulation and energy efficiency and permitting domestic solar and wind energy for each dwelling, to counteract the expected rise in demand for energy when homes are in use during hot and cold daytime hours. Local authorities could promote high-quality local community neighborhoods through good public services, attractive facilities and amenities, green spaces, high walkability, and bicycle paths.

All chapters point to the benefits of localization around neighborhood structures. Experience in China demonstrates how neighborhood-level community structures can function, especially in times of need. That may require a decentralization of decision making, which could be encouraged through participatory budgeting for projects at the neighborhood level and by strengthening the role of community centers with staff and budgets.

In previous chapters, we have noted a potential risk of inequality and even discrimination that could occur at neighborhood level, welcoming "like-minded" and keeping out "non-like-minded". That risk should be of particular concern to local authorities, who could allocate increased budgets, public services, and staff to assist poorer and less attractive neighborhoods.

The threat of deterioration of city centers through loss of functions requires a strategic review by all municipal authorities. Not only will they have lost commercial functions but they will also lose income, since commercial properties paid higher rates than residential properties and working from home in residential areas will not bring a public willingness to pay higher rates. Municipal authorities will be seeking ways to attract millennials and young populations back to city centers, to revive urbanity and city life, cafes and restaurants, and the role of cities as places of creativity, innovation, and spontaneous connectivity. Public-private partnerships will be needed to bring non-work leisure, pleasure, and culture back to life in city centers.

Since circulation patterns may not return to former destinations, cities will need to review where to invest funds in urban transportation projects. If commuter levels fall, high-cost infrastructures may not be justified and the budgets available could be better used for enabling micro-mobility (non-car but may include single/double seat electric vehicles as well as bicycles) in neighborhoods and for access to mass transit nodes.

• Mindful consumption

Mindful consumption could be lost if governments and policy makers do not take the opportunity to support and promote positive trends and scale them up to become mainstream. That will require deliberate orientation of financial support, public purchasing and infrastructure development, provision of public services, and regulation that could facilitate, encourage, or require preference for more sustainable consumption. Some already exist but public's positive attitudes may now enable additional steps to be taken.

Governments act as major consumers, purchasing contracts for supplies to public institutions, in addition to their role as regulators. They can require suppliers to provide products and services to governmental and public institutions, which give preference to a more sustainable choice. Obvious examples are the adoption of green building in contracts for public buildings, buying and using electric means of transportation, or the choice of organic, small family-based agriculture food supplies to public institutions.

However, governments can also discourage more sustainable consumption, by counteracting the more favorable trends that emerged during Covid-19 for a reduction of consumerism and an increase in savings. Indiscriminate financial support to business and individuals to revive shopping as a means to revive the economy misses an opportunity to provide necessary support in a way that could benefit more sustainable consumption. Although emergency assistance in the form of universal basic income is a welcome direction for social sustainability, a preferable form of assistance would be universal basic services, which would cover costs for essential services, such as healthcare, higher education and reskilling, public transportation, and digital connectivity.

Another counteraction is occurring where governments are recalling or removing environmental safeguards on production and consumer products in order to activate the economy. The opposite policy is needed, using public's positive attitudes to promote ethical consumerism, for example, for healthier non-meat diets.

Many countries are considering how they can promote the economy by combining to jump-start the economy post-Covid-19 with investing in public infrastructures. Some are deliberately promoting "green" infrastructures, such as providing financial encouragement for renewable energy (EU Green Deal, Chapter 8). We suggest that public infrastructure investment be oriented to enabling and nudging people to choose and to live more sustainable lifestyles, incorporating mindful consumption. That would include investing directly in infrastructures that influence choice or by providing incentives to encourage uptake of a more sustainable choice. Examples include domestic energy efficiency, use of micro-mobility within neighborhoods, comfortable, convenient, and affordable public transport between neighborhoods, and high-speed rail as an alternative to short distance flights and to car transportation on highways.

- Sharing and counteracting waste

Reducing and preventing waste had progressed pre-Covid-19; relevant policies, strategies, and instruments are well known to all countries and civil society organizations. The problem today is how to restrain the reversal and loss of progress resulting from mass use of disposable products, an increase in waste from packaging and delivery, and a reluctance to share, reuse, or recycle. The extra waste imposed an additional collection and disposal burden on local authorities and in places resulted in a deterioration of cleanliness in city streets and parks.

Packaging will be an ongoing and increasing problem as online shopping expands. This is an opportunity for designers and business enterprises, as well as for regulators to find and mandate more environmentally acceptable materials and solutions to removing the need for unnecessary packaging, finding biodegradable packaging, ensuring recycling of packaging, and obtaining the cooperation of the public to reuse, repair, and recycle.

- Counteracting the car

Although traffic congestion was still a major concern in cities, there had been some progress pre-Covid-19 to attracting commuters to using mass transit and public transport and to leave their cars at home. Millennials chose to live in city centers without a car. Covid-19 has brought the private car back as a "safe" Covid-19-proof form of individual transport, at the expense of mass transit operations. Traffic congestion returned on exiting lockdowns even though fewer people were commuting to work. Electric vehicles only provide a short-term solution diminishing greenhouse gas emissions; but in the long term, we should aim to remove the car altogether from city streets and greatly reduce its use elsewhere.

The key issue is how to counteract the increased use of the private car, persuade the public to return to mass transit and public transport, and continue to enjoy extra city space for non-car activities like walking, biking, and outdoor cafes and restaurants. Strategies, policies, and tools are well known, but in the past their implementation encountered public and political opposition. After enjoying the transformation of roadways and parking areas to bicycle lanes, pedestrian space, and seating areas for cafes and restaurants, and appreciating cleaner air, some measures may now be more acceptable and even welcomed by the public.

Cities could revive proposals for congestion tax, priority lanes for public transport, and a reduction of parking spaces. Investment in infrastructure for micro-mobility, walkability, and bike-ability would help to embed the change of attitudes toward living locally around neighborhoods. Permitting commercial activities to spread out over roadways and parking spaces could enable businesses to flourish.

Last but not least, measures will be needed to reassure the public that travel by mass transit and public transport is not only safer but also better for all.

Outlook

The materials, scenarios, discussions, reflections, and projections presented in this book are written and presented in the middle of an ongoing pandemic, with only a remote outlook on a vaccine that is widely available. Obviously, it is far too early to draw definitive conclusions, especially with respect to sustainability. Climate change and other environmental hazards are not going away; but the current crisis has shown that large-scale lifestyle changes are possible when the necessity is shared at a large scale, and when government steps in with mandates and support mechanisms. Not all lifestyle changes are beneficial for sustainability, but several may be.

The pandemic should be considered a huge collective learning experiment in which governments, business, and civil society are gaining comprehension and skills about how to cope with large, unknown, and unexpected challenges. The outcomes of these learning processes are still unknown. We may expect that preparedness for future epidemics and pandemic will improve, making use of increased cyber literacy among the population. We may hope that the new roles governments have to play will not lead to more authoritarian forms of government, but rather to the involvement of local level communities and governments that have been lost during the neoliberal turn in the 1980s. We may also hope that ongoing crises like climate change will be tackled by newly self-confident governments that are not afraid to spend big money on societal challenges, a lesson learned through the Covid-19 pandemic.

Index

Note: Page numbers in *italics* indicate a figure and page numbers in **bold** indicate a table on the corresponding page.